THE ĀRYABHAṬĪYA
of
ĀRYABHAṬA

THE UNIVERSITY OF CHICAGO PRESS
CHICAGO, ILLINOIS

THE BAKER & TAYLOR COMPANY
NEW YORK

THE CAMBRIDGE UNIVERSITY PRESS
LONDON

THE MARUZEN-KABUSHIKI-KAISHA
TOKYO, OSAKA, KYOTO, FUKUOKA, SENDAI

THE COMMERCIAL PRESS, LIMITED
SHANGHAI

THE ĀRYABHAṬĪYA

of

ĀRYABHAṬA

An Ancient Indian Work on Mathematics and Astronomy

TRANSLATED WITH NOTES BY

WALTER EUGENE CLARK

Professor of Sanskrit in Harvard University

THE UNIVERSITY OF CHICAGO PRESS
CHICAGO, ILLINOIS

COMPOSED AND PRINTED BY THE UNIVERSITY OF CHICAGO PRESS
CHICAGO, ILLINOIS, U.S.A.

PREFACE

In 1874 Kern published at Leiden a text called the *Āryabhaṭīya* which claims to be the work of Āryabhaṭa, and which gives (III, 10) the date of the birth of the author as 476 A.D. If these claims can be substantiated, and if the whole work is genuine, the text is the earliest preserved Indian mathematical and astronomical text bearing the name of an individual author, the earliest Indian text to deal specifically with mathematics, and the earliest preserved astronomical text from the third or scientific period of Indian astronomy. The only other text which might dispute this last claim is the *Sūryasiddhānta* (translated with elaborate notes by Burgess and Whitney in the sixth volume of the *Journal of the American Oriental Society*). The old *Sūryasiddhānta* undoubtedly preceded Āryabhaṭa, but the abstracts from it given early in the sixth century by Varāhamihira in his *Pañcasiddhāntikā* show that the preserved text has undergone considerable revision and may be later than Āryabhaṭa. Of the old *Pauliśa* and *Romaka Siddhāntas*, and of the transitional *Vāsiṣṭha Siddhānta*, nothing has been preserved except the short abstracts given by Varāhamihira. The names of several astronomers who preceded Āryabhaṭa, or who were his contemporaries, are known, but nothing has been preserved from their writings except a few brief fragments.

The *Āryabhaṭīya*, therefore, is of the greatest importance in the history of Indian mathematics and astronomy. The second section, which deals with mathematics (the *Gaṇitapāda*), has been translated by Rodet in the *Journal asiatique* (1879), I, 393–434, and by Kaye in the *Journal of the Asiatic Society of Bengal*, 1908, pages 111–41. Of the rest of the work no translation has appeared, and only a few of the stanzas have been discussed. The aim of this work is to give a complete translation of the *Āryabhaṭīya* with references to some of the most important parallel passages which may be of assistance for further study. The edition of Kern makes no pretense of giving a really critical text of the *Āryabhaṭīya*. It gives merely the text which the sixteenth-century commentator Parameśvara had before him. There are several uncertainties about this text. Especially noteworthy is the considerable gap after IV, 44, which is discussed by Kern (pp. v–vi). The names of other commentators have been noticed by Bibhutibhusan Datta in the *Bulletin of the Calcutta Mathematical Society*, XVIII (1927), 12. All available manuscripts of the text should be consulted, all the other commentators should be studied, and a careful comparison of the *Āryabhaṭīya* with the abstracts from the old *siddhāntas* given by Varāhamihira, with the *Sūryasiddhānta*, with the *Śiṣyadhīvṛddhida* of Lalla, and with the *Brāhmasphuṭasiddhānta* and the *Khaṇḍakhādyaka* of Brahmagupta should be made. All the later quotations from Āryabhaṭa, especially those made by the commentators on Brahmagupta and Bhāskara, should be collected and verified. Some of those noted by Colebrooke do not seem to fit the published *Ārya-*

bhaṭīya. If so, were they based on a lost work of Āryabhaṭa, on the work of another Āryabhaṭa, or were they based on later texts composed by followers of Āryabhaṭa rather than on a work by Āryabhaṭa himself? Especially valuable would be a careful study of Pṛthūdakasvāmin or Caturvedācārya, the eleventh-century commentator on Brahmagupta, who, to judge from Sudhākara's use of him in his edition of the *Brāhmasphuṭasiddhānta*, frequently disagrees with Brahmagupta and upholds Āryabhaṭa against Brahmagupta's criticisms.

The present translation, with its brief notes, makes no pretense at completeness. It is a preliminary study based on inadequate material. Of several passages no translation has been given or only a tentative translation has been suggested. A year's work in India with unpublished manuscript material and the help of competent pundits would be required for the production of an adequate translation. I have thought it better to publish the material as it is rather than to postpone publication for an indefinite period. The present translation will have served its purpose if it succeeds in attracting the attention of Indian scholars to the problem, arousing criticism, and encouraging them to make available more adequate manuscript material.

There has been much discussion as to whether the name of the author should be spelled Āryabhaṭa or Āryabhaṭṭa.[1] *Bhaṭa* means "hireling," "mercenary,"

[1] See especially *Journal of the Royal Asiatic Society*, 1865, pp. 392–93; *Journal asiatique* (1880), II, 473–81; Sudhākara Dvivedī, *Gaṇakataraṅgiṇī*, p. 2.

"warrior," and *bhaṭṭa* means "learned man," "scholar." Āryabhaṭṭa is the spelling which would naturally be expected. However, all the metrical evidence seems to favor the spelling with one *ṭ*. It is claimed by some that the metrical evidence is inclusive, that *bhaṭa* has been substituted for *bhaṭṭa* for purely metrical reasons, and does not prove that Āryabhaṭa is the correct spelling. It is pointed out that Kern gives the name of the commentator whom he edited as Paramādīśvara. The name occurs in this form in a stanza at the beginning of the text and in another at the end, but in the prose colophons at the ends of the first three sections the name is given as Parameśvara, and this doubtless is the correct form. However, until more definite historical or metrical evidence favoring the spelling Āryabhaṭṭa is produced I prefer to keep the form Āryabhaṭa.

The *Āryabhaṭīya* is divided into four sections which contain in all only 123 stanzas. It is not a complete and detailed working manual of mathematics and astronomy. It seems rather to be a brief descriptive work intended to supplement matters and processes which were generally known and agreed upon, to give only the most distinctive features of Āryabhaṭa's own system. Many commonplaces and many simple processes are taken for granted. For instance, there are no rules to indicate the method of calculating the *ahargaṇa* and of finding the mean places of the planets. But rules are given for calculating the true places from the mean places by applying certain corrections, although even here there is no statement of the method by which the corrections themselves are to be calculated. It is a descriptive summary rather than a full working manual like the later *karaṇagranthas* or the *Sūryasiddhānta* in its present form. It is questionable whether Āryabhaṭa himself composed another treatise, a *karaṇagrantha* which might serve directly as a basis for practical calculation, or whether his methods were confined to oral tradition handed down in a school.

Brahmagupta[1] implies knowledge of two works by Āryabhaṭa, one giving three hundred *sāvana* days in a *yuga* more than the other, one beginning the *yuga* at sunrise, the other at midnight. He does not seem to treat these as works of two different Āryabhaṭas. This is corroborated by *Pañcasiddhāntikā*, XV, 20: "Āryabhaṭa maintains that the beginning of the day is to be reckoned from midnight at Laṅkā; and the same teacher [*sa eva*] again says that the day begins from sunrise at Laṅkā." Brahmagupta, however, names only the *Daśagītika* and the *Āryāṣṭaśata* as the works of Āryabhaṭa, and these constitute our *Āryabhaṭīya*. But the word *audayikatantra* of *Brāhmasphuṭasiddhānta*, XI, 21 and the words *audayika* and *ārdharātrika* of XI, 13–14 seem to imply that Brahmagupta is distinguishing between two works of one Āryabhaṭa. The published *Āryabhaṭīya* (I, 2) begins the *yuga* at sunrise. The other work may not have been named or criticized by Brahmagupta because of the fact that it followed orthodox tradition.

Alberuni refers to two Āryabhaṭas. His later

[1] *Brāhmasphuṭasiddhānta*, XI, 5 and 13–14.

Āryabhaṭa (of Kusumapura) cannot be the later Āryabhaṭa who was the author of the *Mahāsiddhānta*. The many quotations given by Alberuni prove conclusively that his second Āryabhaṭa was identical with the author of our *Āryabhaṭīya* (of Kusumapura as stated at II, 1). Either there was a still earlier Āryabhaṭa or Alberuni mistakenly treats the author of our *Āryabhaṭīya* as two persons. If this author really composed two works which represented two slightly different points of view it is easy to explain Alberuni's mistake.[1]

The published text begins with 13 stanzas, 10 of which give in a peculiar alphabetical notation and in a very condensed form the most important numerical elements of Āryabhaṭa's system of astronomy. In ordinary language or in numerical words the material would have occupied at least four times as many stanzas. This section is named *Daśagītikasūtra* in the concluding stanza of the section. This final stanza, which is a sort of colophon; the first stanza, which is an invocation and which states the name of the author; and a *paribhāṣā* stanza, which explains the peculiar alphabetical notation which is to be employed in the following 10 stanzas, are not counted. I see nothing suspicious in the discrepancy as Kaye does. There is no more reason for questioning the authenticity of the *paribhāṣā* stanza than for questioning that of the invocation and colophon. Kaye

[1] For a discussion of the whole problem of the two or three Āryabhaṭas see Kaye, *Bibliotheca mathematica*, X, 289, and Bibhutibhusan Datta, *Bulletin of the Calcutta Mathematical Society*, XVII (1926), 59.

would like to eliminate it since it seems to furnish evidence for Āryabhaṭa's knowledge of place-value. Nothing is gained by doing so since Lalla gives in numerical words the most important numerical elements of Āryabhaṭa without change, and even without this *paribhāṣā* stanza the rationale of the alphabetical notation in general could be worked out and just as satisfactory evidence of place-value furnished. Further, Brahmagupta (*Brāhmasphuṭasiddhānta*, XI, 8) names the *Daśagītika* as the work of Āryabhaṭa, gives direct quotations (XI, 5; I, 12 and XI, 4; XI, 17) of stanzas 1, 3, and 4 of our *Daśagītika*, and XI, 15 (although corrupt) almost certainly contains a quotation of stanza 5 of our *Daśagītika*. Other stanzas are clearly referred to but without direct quotations. Most of the *Daśagītika* as we have it can be proved to be earlier than Brahmagupta (628 A.D.).

The second section in 33 stanzas deals with mathematics. The third section in 25 stanzas is called *Kālakriyā*, or "The Reckoning of Time." The fourth section in 50 stanzas is called *Gola*, or "The Sphere." Together they contain 108 stanzas.

The *Brāhmasputasiddhānta* of Brahmagupta was composed in 628 A.D., just 129 years after the *Āryabhaṭīya*, if we accept 499 A.D., the date given in III, 10, as being actually the date of composition of that work. The eleventh chapter of the *Brāhmasphuṭasiddhānta*, which is called "Tantraparīkṣā," and is devoted to severe criticism of previous works on astronomy, is chiefly devoted to criticism of Āryabhaṭa. In this chapter, and in other parts of his work,

Brahmagupta refers to Āryabhaṭa some sixty times. Most of these passages contain very general criticism of Āryabhaṭa as departing from *smṛti* or being ignorant of astronomy, but for some 30 stanzas it can be shown that the identical stanzas or stanzas of identical content were known to Brahmagupta and ascribed to Āryabhaṭa. In XI, 8 Brahmagupta names the *Āryāṣṭaśata* as the work of Āryabhaṭa, and XI, 43, *jānāty ekam api yato nāryabhaṭo gaṇitakālagolānām,* seems to refer to the three sections of our *Āryāṣṭaśata.* These three sections contain exactly 108 stanzas. No stanza from the section on mathematics has been quoted or criticized by Brahmagupta, but it is hazardous to deduce from that, as Kaye does,[1] that this section on mathematics is spurious and is a much later addition.[2] To satisfy the conditions demanded by Brahmagupta's name *Āryāṣṭaśata* there must have been in the work of Āryabhaṭa known to him exactly 33 other stanzas forming a more primitive and less developed mathematics, or these 33 other stanzas must have been astronomical in character, either forming a separate chapter or scattered through the present third and fourth sections. This seems to be most unlikely. I doubt the validity of Kaye's contention that the *Gaṇitapāda* was later than Brahmagupta. His suggestion that it is by the later Āryabhaṭa who was the author of the *Mahāsiddhānta* (published in the "Benares Sanskrit

[1] *Op. cit.*, X, 291–92.

[2] For criticism of Kaye see Bibhutibhusan Datta, *op. cit.*, XVIII (1927), 5.

Series" and to be ascribed to the tenth century or even later) is impossible, as a comparison of the two texts would have shown.

I feel justified in assuming that the *Āryabhaṭīya* on the whole is genuine. It is, of course, possible that at a later period some few stanzas may have been changed in wording or even supplanted by other stanzas. Noteworthy is I, 4, of which the true reading *bhūḥ*, as preserved in a quotation of Brahmagupta, has been changed by Parameśvara or by some preceding commentator to *bham* in order to eliminate Āryabhaṭa's theory of the rotation of the Earth.

Brahmagupta criticizes some astronomical matters in which Āryabhaṭa is wrong or in regard to which Āryabhaṭa's method differs from his own, but his bitterest and most frequent criticisms are directed against points in which Āryabhaṭa was an innovator and differed from *smṛti* or tradition. Such criticism would not arise in regard to mathematical matters which had nothing to do with theological tradition. The silence of Brahmagupta here may merely indicate that he found nothing to criticize or thought criticism unnecessary. Noteworthy is the fact that Brahmagupta does not give rules for the volume of a pyramid and for the volume of a sphere, which are both given incorrectly by Āryabhaṭa (II, 6–7). This is as likely to prove ignorance of the true values on Brahmagupta's part as lateness of the rules of Āryabhaṭa. What other rules of the *Gaṇitapāda* could be open to adverse criticism? On the positive side may be pointed out the very close correspondence in terminology and expression between the fuller text of Brahmagupta, XVIII, 3–5 and the more enigmatical text of *Āryabhaṭīya*, II, 32–33, in their statements of the famous Indian method (*kuṭṭaka*) of solving indeterminate equations of the first degree. It seems probable to me that Brahmagupta had before him these two stanzas in their present form. It must be left to the mathematicians to decide which of the two rules is earlier.

The only serious internal discrepancy which I have been able to discover in the *Āryabhaṭīya* is the following. Indian astronomy, in general, maintains that the Earth is stationary and that the heavenly bodies revolve about it, but there is evidence in the *Āryabhaṭīya* itself and in the accounts of Āryabhaṭa given by later writers to prove that Āryabhaṭa maintained that the Earth, which is situated in the center of space, revolves on its axis, and that the asterisms are stationary. Later writers attack him bitterly on this point. Even most of his own followers, notably Lalla, refused to follow him in this matter and reverted to the common Indian tradition. Stanza IV, 9, in spite of Parameśvara, must be interpreted as maintaining that the asterisms are stationary and that the Earth revolves. And yet the very next stanza (IV, 10) seems to describe a stationary Earth around which the asterisms revolve. Quotations by Bhaṭṭotpala, the Vāsanāvārttika, and the Marīci indicate that this stanza was known in its present form from the eleventh century on. Is it capable of some different interpretation? Is it intended merely as a statement of the popular view? Has its wording been changed as has been done with I, 4? I see at present no satisfactory solution of the problem.

Colebrooke[1] gives *caturviṁśaty aṁśaiś cakram ubhayato gacchet* as a quotation by Munīśvara from the *Āryāṣṭaśata* of Āryabhaṭa. This would indicate a knowledge of a libration of the equinoxes. No such statement is found in our *Āryāṣṭaśata*. The quotation should be verified in the unpublished text in order to determine whether Colebrooke was mistaken or whether we are faced by a real discrepancy. The words are not found in the part of the Marīci which has already been published in the *Pandit*.

The following problem also needs elucidation. Although Brahmagupta (XI, 43–44)

> jānāty ekam api yato nāryabhaṭo gaṇitakālagolānām |
> na mayā proktāni tataḥ pṛthak pṛthag dūṣaṇāny eṣām ||
> āryabhaṭadūṣaṇānāṁ saṁkhyā vaktuṁ na śakyate yasmāt |
> tasmād ayam uddeśo buddhimatānyāni yojyāni ||

sums up his criticism of Āryabhaṭa in the severest possible way, yet at the beginning of his *Khaṇḍakhādyaka*, a *karaṇagrantha* which has recently been edited by Babua Misra Jyotishacharyya (University of Calcutta, 1925), we find the statement *vakṣyāmi khaṇḍakhādyakam ācāryāryabhaṭatulyaphalam*. It is curious that Brahmagupta in his *Khaṇḍakhādyaka* should use such respectful language and should follow the authority of an author who was damned so unmercifully by him in the *Tantraparīkṣā* of his *Brāhmasphuṭasiddhānta*. Moreover, the elements of the *Khaṇ-*

[1] *Miscellaneous Essays*, II, 378.

ḍakhādyaka seem to differ much from those of the *Āryabhaṭīya.*[1] Is this to be taken as an indication that Brahmagupta here is following an older and a different Āryabhaṭa? If so the *Brāhmasphuṭasiddhānta* gives no clear indication of the fact. Or is he following another work by the same Āryabhaṭa? According to Dīkṣit,[2] the *Khaṇḍakhādyaka* agrees in all essentials with the old form of the *Sūryasiddhānta* rather than with the *Brāhmasphuṭasiddhānta.* Just as Brahmagupta composed two different works so Āryabhaṭa may have composed two works which represented two different points of view. The second work may have been cast in a traditional mold, may have been based on the old *Sūryasiddhānta,* or have formed a commentary upon it.

The *Mahāsiddhānta* of another Āryabhaṭa who lived in the tenth century or later declares (XIII, 14):

vṛddhāryabhaṭaproktāt siddhāntād yan mahākālāt |
pāṭhair gatam ucchedaṁ viśeṣitaṁ tan mayā svoktyā ||

But this *Mahāsiddhānta* differs in so many particulars from the *Āryabhaṭīya* that it is difficult to believe that the author of the *Āryabhaṭīya* can be the one referred to as Vṛddhāryabhaṭa unless he had composed another work which differed in many particulars from the *Āryabhaṭīya.* The matter needs careful investigation.[3]

[1] Cf. *Pañcasiddhāntikā,* p. xx, and *Bulletin of the Calcutta Mathematical Society,* XVII (1926), 69.

[2] As reported by Thibaut, *Astronomie, Astrologie und Mathematik,* pp. 55, 59.

[3] See *Bulletin of the Calcutta Mathematical Society,* XVII (1926), 66–67, for a brief discussion.

This monograph is based upon work done with me at the University of Chicago some five years ago by Baidyanath Sastri for the degree of A.M. So much additional material has been added, so many changes have been made, and so many of the views expressed would be unacceptable to him that I have not felt justified in placing his name, too, upon the title-page as joint-author and thereby making him responsible for many things of which he might not approve.

Harvard University
April, 1929

While reading the final page-proof I learned of the publication by Prabodh Chandra Sengupta of a translation of the Āryabhaṭīya in the *Journal of the Department of Letters* (Calcutta University), XVI (1927). Unfortunately it has not been possible to make use of it in the present publication.

April, 1930

TABLE OF CONTENTS

LIST OF ABBREVIATIONS

Alberuni.................... Alberuni's *India*. Translated by E. C. Sachau. London, 1910.

Barth (*Œuvres*)............. *Œuvres de Auguste Barthe.* 3 vols. Paris, 1917.

BCMS..................... *Bulletin of the Calcutta Mathematical Society.*

Bhāskara, *Gaṇitādhyāya*..... Edited by Bapu Deva Sastri; revised by Ramachandra Gupta. Benares (no date).

Bhāskara, *Golādhyāya*....... Edited by Bapu Deva Sastri; revised by Ramachandra Gupta. Benares (no date).
Edited by Girija Prasad Dvivedi. Lucknow: Newul Kishore Press, 1911.

Bhaṭṭotpala................ The *Bṛhat Saṁhitā* by Varāhamihira with the commentary of Bhaṭṭotpala. "Vizianagram Sanskrit Series," Vol. X. Benares, 1895–97.

Bibl. math................. *Bibliotheca mathematica.*

Brahmagupta.............. Refers to *Brāhmasphuṭasiddhānta.*

Brāhmasphuṭasiddhānta...... Edited by Sudhākara Dvivedin in the *Pandit* (N.S.), Vols. XXIII–XXIV. Benares, 1901–2.

Brennand, *Hindu Astronomy*.. W. Brennand, *Hindu Astronomy*. London, 1896.

Bṛhat Saṁhitā.............. The *Bṛhat Saṁhitā* by Varāhamihira with the commentary of Bhaṭṭotpala. "Vizianagram Sanskrit Series," Vol. X. Benares, 1895–97.

Colebrooke, *Algebra*	H. T. Colebrooke, *Algebra, with Arithmetic and Mensuration, from the Sanscrit of Brahmegupta and Bháscara.* London, 1817.
Colebrooke, *Essays*	*Miscellaneous Essays* (2d ed.), by H. T. Colebrooke. Madras, 1872.
Hemacandra, *Abhidhāna-cintāmaṇi*	Edited by Böhtlingk and Rieu. St. Petersburg, 1847.
IA	*Indian Antiquary.*
IHQ	*Indian Historical Quarterly.*
JA	*Journal asiatique.*
JASB	*Journal and Proceedings of the Asiatic Society of Bengal.*
JBBRAS	*Journal of the Bombay Branch of the Royal Asiatic Society.*
JBORS	*Journal of the Bihar and Orissa Research Society.*
JIMS	*Journal of the Indian Mathematical Society.*
JRAS	*Journal of the Royal Asiatic Society.*
Kaye, *Indian Mathematics*	G. R. Kaye, *Indian Mathematics.* Calcutta, 1915.
Kaye, *Hindu Astronomy*	"Memoirs of the Archaeological Survey of India," No. 18. Calcutta, 1924.
Khaṇḍakhādyaka	By Brahmagupta. Edited by Babua Misra Jyotishacharyya. University of Calcutta, 1925.
Lalla	The *Śiṣyadhīvṛddhida* of Lalla. Edited by Sudhākara Dvivedin. Benares (no date).
Mahāsiddhānta	By Āryabhaṭa. Edited by Sudhākara Dvivedin in the "Benares Sanskrit Series." 1910.

Marīci . . . . The *Gaṇitādhyāya* of Bhāskara's *Siddhāntaśiromaṇi* with Vāsanābhāṣya, Vāsanāvārttika, and Marīci. *Pandit* (N.S.), Vols. XXX–XXXI. Benares, 1908–9.

Pañcasiddhāntikā . . . . G. Thibaut and Sudhākara Dvivedī, *The Pañcasiddhāntikā. The Astronomical Work of Varāha Mihira*. Benares, 1889.

Sudhākara, *Gaṇakataraṅgiṇī*. . Benares, 1892.

Sūryasiddhānta . . . . Edited by F. E. Hall and Bapu Deva Sastrin in the *Bibliotheca indica*. Calcutta, 1859.

Translated by Burgess and Whitney, *Journal of the American Oriental Society*, Vol. VI.

Vāsanāvārttika . . . . The *Gaṇitādhyāya* of Bhāskara's *Siddhāntaśiromaṇi* with Vāsanābhāṣya, Vāsanāvārttika, and Marīci. *Pandit* (N.S.), Vols. XXX–XXXI. Benares, 1908–9.

ZDMG . . . . *Zeitschrift der Deutschen Morgenländischen Gesellschaft*.

I, II, III, and IV refer to the four sections of the *Āryabhaṭīya*.

CHAPTER I

DAŚAGĪTIKA OR THE TEN GĪTI STANZAS

A. Having paid reverence to Brahman, who is one (in causality, as the creator of the universe, but) many (in his manifestations), the true deity, the Supreme Spirit, Āryabhaṭa sets forth three things: mathematics [*gaṇita*], the reckoning of time [*kālakriyā*], and the sphere [*gola*].

Baidyanath suggests that *satyā devatā* may denote Sarasvatī, the goddess of learning. For this I can find no support, and therefore follow the commentator Parameśvara in translating "the true deity," God in the highest sense of the word, as referring to Prajāpati, Pitāmaha, Svayambhū, the lower individualized Brahman, who is so called as being the creator of the universe and above all the other gods. Then this lower Brahman is identified with the higher Brahman as being only an individualized manifestation of the latter. As Parameśvara remarks, the use of the word *kam* seems to indicate that Āryabhaṭa based his work on the old *Pitāmahasiddhānta*. Support for this view is found in the concluding stanza of our text (IV, 50), *āryabhaṭīyaṁ nāmnā pūrvaṁ svāyambhuvaṁ sadā sad yat*. However, as shown by Thibaut[1] and Kharegat,[2] there is a close connection between Āryabhaṭa and the old *Sūryasiddhānta*. At

[1] *Pañcasiddhāntikā*, pp. xviii, xxvii.

[2] *JBBRAS*, XIX, 129–31.

present the evidence is too scanty to allow us to specify the sources from which Āryabhaṭa drew.

The stanza has been translated by Fleet.[1] As pointed out first by Bhāu Dājī,[2] a passage of Brahmagupta (XII, 43), *jānāty ekam api yato nāryabhaṭo gaṇitakālagolānām*, seems to refer to the *Gaṇitapāda*, the *Kālakriyāpāda*, and the *Golapāda* of our *Āryabhaṭīya* (see also Bibhutibhusan Datta).[3] Since Brahmagupta (XI, 8) names the *Daśagītika* and the *Āryāṣṭaśata* (108 stanzas) as works of Āryabhaṭa, and since the three words of XI, 43 refer in order to the last three sections of the *Āryabhaṭīya* (which contain exactly 108 stanzas), their occurrence there in this order seems to be due to more than mere coincidence. As Fleet remarks,[4] Āryabhaṭa here claims specifically as his work only three chapters. But Brahmagupta (628 A.D.) actually quotes at least three passages of our *Daśagītika* and ascribes it to Āryabhaṭa. There is no good reason for refusing to accept it as part of Āryabhaṭa's treatise.

B. Beginning with *ka* the *varga* letters (are to be used) in the *varga* places, and the *avarga* letters (are to be used) in the *avarga* places. *Ya* is equal to the sum of *ṅa* and *ma*. The nine vowels (are to be used) in two nines of places *varga* and *avarga*. *Navāntyavarge vā*.

Āryabhaṭa's system of expressing numbers by means of letters has been discussed by Whish,[5] by

[1] *JRAS*, 1911, pp. 114–15.

[2] *Ibid.*, 1865, p. 403.

[3] *BCMS*, XVIII (1927), 16.

[4] *JRAS*, 1911, pp. 115, 125.

[5] *Transactions of the Literary Society of Madras*, I (1827), 54, translated with additional notes by Jacquet, *JA* (1835), II, 118.

Brockhaus,[1] by Kern,[2] by Barth,[3] by Rodet,[4] by Kaye,[5] by Fleet,[6] by Sarada Kanta Ganguly,[7] and by Sukumar Ranjan Das.[8] I have not had access to the *Pṛthivīr Itihāsa* of Durgadas Lahiri.[9]

The words *varga* and *avarga* seem to refer to the Indian method of extracting the square root, which is described in detail by Rodet[10] and by Avadhesh Narayan Singh.[11] I cannot agree with Kaye's statement[12] that the rules given by Āryabhaṭa for the extraction of square and cube roots (II, 4–5) "are perfectly general (i.e., algebraical)" and apply to all arithmetical notations, nor with his criticism of the foregoing stanza: "Usually the texts give a verse explaining this notation, but this explanatory verse is not Āryabhaṭa's."[13] Sufficient evidence has not been adduced by him to prove either assertion.

The *varga* or "square" places are the first, third, fifth, etc., counting from the right. The *avarga* or "non-square" places are the second, fourth, sixth, etc., counting from the right. The words *varga* and *avarga* seem to be used in this sense in II, 4. There is no good reason for refusing to take them in the same sense here. As applied to the Sanskrit alphabet the *varga* letters referred to here are those from *k* to *m*,

1 *Zeitschrift für die Kunde des Morgenlandes*, IV, 81.
2 *JRAS*, 1863, p. 380.
3 *Œuvres*, III, 182.
4 *JA* (1880), II, 440.
5 *JASB*, 1907, p. 478.
6 *Op. cit.*, 1911, p. 109.
7 *BCMS*, XVII (1926), 195.
8 *IHQ*, III, 110.
9 III, 332 ff.
10 *Op. cit.* (1879), I, 406–8.
11 *BCMS*, XVIII (1927), 128
12 *Op. cit.*, 1908, p. 120.
13 *Ibid.*, p. 118.

which are arranged in five groups of five letters each. The *avarga* letters are those from *y* to *h*, which are not so arranged in groups. The phrase "beginning with *ka*" is necessary because the vowels also are divided into *vargas* or "groups."

Therefore the vowel *a* used in *varga* and *avarga* places with *varga* and *avarga* letters refers the *varga* letters *k* to *m* to the first *varga* place, the unit place, multiplies them by 1. The vowel *a* used with the *avarga* letters *y* to *h* refers them to the first *avarga* place, the place of ten's, multiplies them by 10. In like manner the vowel *i* refers the letters *k* to *m* to the second *varga* place, the place of hundred's, multiplies them by 100. It refers the *avarga* letters *y* to *h* to the second *avarga* place, the place of thousand's, multiplies them by 1,000. And so on with the other seven vowels up to the ninth *varga* and *avarga* places. From Āryabhaṭa's usage it is clear that the vowels to be employed are *a*, *i*, *u*, *ṛ*, *ḷ*, *e*, *ai*, *o*, and *au*. No distinction is made between long and short vowels.

From Āryabhaṭa's usage it is clear that the letters *k* to *m* have the values of 1–25. The letters *y* to *h* would have the values of 3–10, but since a short *a* is regarded as inherent in a consonant when no other vowel sign is attached and when the *virāma* is not used, and since short *a* refers the *avarga* letters to the place of ten's, the signs *ya*, etc., really have the values of 30–100.[1] The vowels themselves have no numerical values. They merely serve to refer the consonants (which do have numerical values) to certain places.

[1] See Sarada Kanta Ganguly, *op. cit.*, XVII (1926), 202.

The last clause, which has been left untranslated, offers great difficulty. The commentator Parameśvara takes it as affording a method of expressing still higher numbers by attaching *anusvāra* or *visarga* to the vowels and using them in nine further *varga* (and *avarga*) places. It is doubtful whether the word *avarga* can be so supplied in the compound. Fleet would translate "in the *varga* place after the nine" as giving directions for referring a consonant to the nineteenth place. In view of the fact that the plural subject must carry over into this clause Fleet's interpretation seems to be impossible. Fleet suggests as an alternate interpretation the emendation of *vā* to *hau*. But, as explained above, *au* refers *h* to the eighteenth place. It would run to nineteen places only when expressed in digits. There is no reason why such a statement should be made in the rule. Rodet translates (without rendering the word *nava*), "(séparément) ou à un groupe terminé par un *varga*." That is to say, the clause has nothing to do with the expression of numbers beyond the eighteenth place, but merely states that the vowels may be attached to the consonants singly as *gara* or to a group of consonants as *gra*, in which latter case it is to be understood as applying to each consonant in the group. So *giri* or *gri* and *guru* or *gru*. Such, indeed, is Āryabhaṭa's usage, and such a statement is really necessary in order to avoid ambiguity, but the words do not seem to warrant the translation given by Rodet. If the words can mean "at the end of a group," and if *nava* can be taken with what precedes, Rodet's interpretation is acceptable. However, I know no other passage which would warrant such a translation of *antyavarge*.

Sarada Kanta Ganguly translates, "[Those] nine [vowels] [should be used] in higher places in a similar manner." It is possible for *vā* to have the sense of "beliebig," "fakultativ," and for *nava* to be separated from *antyavarge*, but the regular meaning of *antya* is "the last." It has the sense of "the following" only at the end of a compound, and the dictionary gives only one example of that usage. If *navāntyavarge* is to be taken as a compound, the translation "in the group following the nine" is all right. But Ganguly's translation of *antyavarge* can be maintained only if he produces evidence to prove that *antya* at the beginning of a compound can mean "the following."

If *nava* is to be separated from *antyavarge* it is possible to take it with what precedes and to translate, "The vowels (are to be used) in two nine's of places, nine in *varga* places and nine in *avarga* places," but *antyavarge vā* remains enigmatical.

The translation must remain uncertain until further evidence bearing on the meaning of *antya* can be produced. Whatever the meaning may be, the passage is of no consequence for the numbers actually dealt with by Āryabhaṭa in this treatise. The largest number used by Āryabhaṭa himself (I, 1) runs to only ten places.

Rodet, Barth, and some others would translate "in the two nine's of zero's," instead of "in the two nine's of places." That is to say, each vowel would serve to add two zero's to the numerical value of the consonant. This, of course, will work from the vowel *i* on, but the vowel *a* does not add two zero's. It adds no zero's or one zero depending on whether it is used with *varga* or *avarga* letters. The fact that *khadvinavake* is amplified by *varge 'varge* is an added difficulty to the translation "zero." It seems to me, therefore, preferable to take the word *kha* in the sense of "space" or better "place."[1] Later the word *kha* is one of the commonest words for "zero," but it is still disputed whether a symbol for zero was actually in use in Āryabhaṭa's time. It is possible that computation may have been made on a board ruled into columns. Only nine symbols may have been in use and a blank column may have served to represent zero.

There is no evidence to indicate the way in which the actual calculations were made, but it seems certain to me that Āryabhaṭa could write a number in signs which had no absolutely fixed values in themselves but which had value depending on the places occupied by them (mounting by powers of 10). Compare II, 2, where in giving the names of classes of numbers he uses the expression *sthānāt sthānaṁ daśaguṇaṁ syāt*, "from place to place each is ten times the preceding."

There is nothing to prove that the actual calculation was made by means of these letters. It is probable that Āryabhaṭa was not inventing a numerical notation to be used in calculation but was devising a system by means of which he might express large,

[1] Cf. Fleet, *op. cit.*, 1911, p. 116.

unwieldy numbers in verse in a very brief form.[1] The alphabetical notation is employed only in the *Daśagītika*. In other parts of the treatise, where only a few numbers of small size occur, the ordinary words which denote the numbers are employed.

As an illustration of Āryabhaṭa's alphabetical notation take the number of the revolutions of the Moon in a *yuga* (I, 1), which is expressed by the word *cayagiyiṅuśuchlṛ*. Taken syllable by syllable this gives the numbers 6 and 30 and 300 and 3,000 and 50,000 and 700,000 and 7,000,000 and 50,000,000. That is to say, 57,753,336. It happens here that the digits are given in order from right to left, but they may be given in reverse order or in any order which will make the syllables fit into the meter. It is hard to believe that such a descriptive alphabetical notation was not based on a place-value notation.

This stanza, as being a technical *paribhāṣā* stanza which indicates the system of notation employed in the *Daśagītika*, is not counted. The invocation and the colophon are not counted. There is no good reason why the thirteen stanzas should not have been named *Daśagītika* (as they are named by Āryabhaṭa himself in stanza C) from the ten central stanzas in Gīti meter which give the astronomical elements of the system. The discrepancy offers no firm support to the contention of Kaye that this stanza is a later addition. The manuscript referred to by Kaye[2] as containing fifteen instead of thirteen stanzas is doubtless com-

[1] See *JA* (1880), II, 454, and *BCMS*, XVII (1926), 201.

[2] *Op. cit.*, 1908, p. 111.

parable to the one referred to by Bhāu Dājī[1] as having two introductory stanzas "evidently an after-addition, and not in the Āryā metre."

1. In a *yuga* the revolutions of the Sun are 4,320,000, of the Moon 57,753,336, of the Earth eastward 1,582,237,500, of Saturn 146,564, of Jupiter 364,224, of Mars 2,296,824, of Mercury and Venus the same as those of the Sun.

2. of the apsis of the Moon 488,219, of (the conjunction of) Mercury 17,937,020, of (the conjunction of) Venus 7,022,388, of (the conjunctions of) the others the same as those of the Sun, of the node of the Moon westward 232,226 starting at the beginning of Meṣa at sunrise on Wednesday at Laṅkā.

The so-called revolutions of the Earth seem to refer to the rotation of the Earth on its axis. The number given corresponds to the number of sidereal days usually reckoned in a *yuga*. Parameśvara, who follows the normal tradition of Indian astronomy and believes that the Earth is stationary, tries to prove that here and in IV, 9 (which he quotes) Āryabhaṭa does not really mean to say that the Earth rotates. His effort to bring Āryabhaṭa into agreement with the views of most other Indian astronomers seems to be misguided ingenuity. There is no warrant for treating the revolutions of the Earth given here as based on false knowledge (*mithyājñāna*), which causes the Earth to seem to move eastward because of the actual westward movement of the planets (see note to I, 4).

In stanza 1 the syllable *ṣu* in the phrase which gives the revolutions of the Earth is a misprint for *bu* as given correctly in the commentary.[2]

[1] *Ibid.*, 1865, p. 397. [2] See *ibid.*, 1911, p. 122 n.

Here and elsewhere in the *Daśagītika* words are used in their stem form without declensional endings.

Lalla (*Madhyamādhikāra*, 3–6, 8) gives the same numbers for the revolutions of the planets, and differs only in giving "revolutions of the asterisms" instead of "revolutions of the Earth."

The *Sūryasiddhānta* (I, 29–34) shows slight variations (see *Pañcasiddhāntikā*, pp. xviii–xix, and Kharegat[1] for the closer relationship of Āryabhaṭa to the old *Sūryasiddhānta*).

Bibhutibhusan Datta,[2] in criticism of the number of revolutions of the planets reported by Alberuni (II, 16–19), remarks that the numbers given for the revolutions of Venus and Mercury really refer to the revolutions of their apsides. It would be more accurate to say "conjunctions."

Alberuni (I, 370, 377) quotes from a book of Brahmagupta's which he calls *Critical Research on the Basis of the Canons* a number for the civil days according to Āryabhaṭa. This corresponds to the number of sidereal days given above (cf. the number of sidereal days given by Brahmagupta [I, 22]).

Compare the figures for the number of revolutions of the planets given by Brahmagupta (I, 15–21) which differ in detail and include figures for the revolutions of the apsides and nodes. Brahmagupta (I, 61)

akṛtāryabhaṭaḥ śīghragam indūccaṁ pātam alpagaṁ svagateḥ |
tithyantagrahaṇānāṁ ghuṇākṣaraṁ tasya saṁvādaḥ ||

criticizes the numbers given by Āryabhaṭa for the revolutions of the apsis and node of the Moon.[3]

[1] *JBBRAS*, XVIII, 129–31. [2] *BCMS*, XVII (1926), 71.

[3] See further Bragmagupta (V, 25) and Alberuni (I, 376).

Brahmagupta (II, 46–47) remarks that according to Āryabhaṭa all the planets were not at the first point of Meṣa at the beginning of the *yuga*. I do not know on what evidence this criticism is based.[1]

Brahmagupta (XI, 8) remarks that according to the *Āryāṣṭaśata* the nodes move while according to the *Daśagītika* the nodes (excepting that of the Moon) are fixed:

āryāṣṭaśate pātā bhramanti daśagītike sthirāḥ pātāḥ |
muktvendupātam apamaṇḍale bhramanti sthirā nātaḥ. ||

This refers to I, 2 and IV, 2. Āryabhaṭa (I, 7) gives the location, at the time his work was composed, of the apsides and nodes of all the planets, and (I, 7 and IV, 2) implies a knowledge of their motion. But he gives figures only for the apsis and node of the Moon. This may be due to the fact that the numbers are so small that he thought them negligible for his purpose.

Brahmagupta (XI, 5) quotes stanza 1 of our text:

yugaravibhagaṇāḥ khyughriti yat proktam tat tayor yugaṁ spaṣṭaṁ |
triśatī ravyudayānāṁ tadantaraṁ hetunā kena. ||[2]

[1] See *Sūryasiddhānta*, pp. 27–28, and *JRAS*, 1911, p. 494.

[2] Cf. *JRAS*, 1865, p. 401. This implies, as Sudhākara says, that Brahmagupta knew two works by Āryabhaṭa each giving the revolutions of the Sun as 4,320,000 but one reckoning 300 *sāvana* days more than the other. Cf. Kharegat (*op. cit.*, XIX, 130). Is the reference to another book by the author of our treatise or was there another earlier Āryabhaṭa? Brahmagupta (XI, 13–14) further implies that he knew two works by an author named Āryabhaṭa in one of which the *yuga* began at sunrise, in the other at midnight (see *JRAS*, 1863, p. 384; *JBBRAS*, XIX, 130–31; *JRAS*, 1911, p. 494; *IHQ*, IV, 506). At any rate, Brahmagupta does not imply knowledge of a second Āryabhaṭa. For the whole problem of the two or three Āryabhaṭas see Kaye (*Bibl. math.*, X, 289) and Bibhutibhusan Datta

3. There are 14 Manus in a day of Brahman [a *kalpa*], and 72 *yugas* constitute the period of a Manu. Since the beginning of this *kalpa* up to the Thursday of the Bhārata battle 6 Manus, 27 *yugas*, and 3 *yugapādas* have elapsed.

The word *yugapāda* seems to indicate that Āryabhaṭa divided the *yuga* into four equal quarters. There is no direct statement to this effect, but also there is no reference to the traditional method of dividing the *yuga* into four parts in the proportion of 4, 3, 2, and 1. Brahmagupta and later tradition ascribes to Āryabhaṭa the division of the *yuga* into four equal parts. For the traditional division see *Sūryasiddhānta* (I, 18–20, 22–23) and Brahmagupta (I, 7–8). For discussion of this and the supposed divisions of Āryabhaṭa see Fleet.[1] Compare III, 10, which gives data for the calculation of the date of the composition of Āryabhaṭa's treatise. It is clear that the fixed point was the beginning of Āryabhaṭa's fourth *yugapāda* (the later Kaliyuga) at the time of the great Bhārata battle in 3102 B.C.

Compare Brahmagupta (I, 9)

yugapādān āryabhaṭaś catvāri samāni kṛtayugādīni |
yad abhihitavān na teṣāṁ smṛtyuktasamānam ekam api ||

and XI, 4

āryabhaṭo yugapādāṁs trīn yātān āha kaliyugādau yat |
tasya kṛtāntar yasmāt svayugādyantau na tat tasmāt ||

(*op. cit.*, XVII [1926], 60–74). The *Pañcasiddhāntikā* also (XV, 20), "Āryabhaṭa maintains that the beginning of the day is to be reckoned from midnight at Laṅkā; and the same teacher again says that the day begins from sunrise at Laṅkā," ascribes the two theories to one Āryabhaṭa.

[1] *Op. cit.*, 1911, pp. 111, 486.

with the commentary of Sudhākara. Brahmagupta (I, 12) quotes stanza I, 3,

manusandhiṁ yugam icchaty āryabhaṭas tanmanur yataḥ śkhayugaḥ |
kalpaś caturyugānāṁ sahasram aṣṭādhikaṁ tasya. ||[1]

Brahmagupta (I, 28) refers to the same matter,

adhikaḥ smṛtyuktamanor āryabhaṭoktaś caturyugeṇa manuḥ |
adhikaṁ viṁśāṁśayutais tribhir yugais tasya kalpagatam. ||

Brahmagupta (XI, 11) criticizes Āryabhaṭa for beginning the Kaliyuga with Thursday (see the commentary of Sudhākara).

Bhāu Dājī[2] first pointed out the parallels in Brahmagupta I, 9 and XI, 4 and XI, 11.[3]

4. The revolutions of the Moon (in a *yuga*) multiplied by 12 are signs [*rāśi*].[4] The signs multiplied by 30 are degrees. The degrees multiplied by 60 are minutes. The minutes multiplied by 10 are *yojanas* (of the circumference of the sky). The Earth moves one minute in a *prāṇa*.[5] The circumference of the sky (in *yojanas*) divided by the revolutions of a planet in a *yuga* gives the *yojanas* of the planet's orbit. The orbit of the Sun is a sixtieth part of the circle of the asterisms.

In translating the words *śaśirāśayaṣ ṭha cakram* I have followed Parameśvara's interpretation *śaśinaś cakraṁ bhagaṇā dvādaśaguṇitā rāśayaḥ*. The Sanskrit construction is a harsh one, but there is no other way of making sense. *Śaśi* (without declensional ending) is to be separated.

Parameśvara explains the word *grahajavo* as fol-

[1] Cf. III, 8. [2] *Op. cit.*, 1865, pp. 400–401.

[3] Cf. Alberuni, I, 370, 373–74.

[4] A *rāśi* is a sign of the zodiac or one-twelfth of a circle.

[5] For *prāṇa* see III, 2.

lows: *ekaparivṛttau grahasya javo gatimānaṁ yojanātmakaṁ bhavati.*

The word *yojanāni* must be taken as given a figure in *yojanas* for the circumference of the sky (*ākāśakakṣyā*). It works out as 12,474,720,576,000, which is the exact figure given by Lalla (*Madhyamādhikāra* 13) who was a follower of Āryabhaṭa. Compare *Sūryasiddhānta*, XII, 80–82; Brahmagupta, XXI, 11–12; Bhāskara, *Golādhyāya, Bhuvanakośa*, 67–69 and *Gaṇitādhyāya, Kakṣādhyāya*, 1–5.

The statement of Alberuni (I, 225) with regard to the followers of Āryabhaṭa,

> It is sufficient for us to know the space which is reached by the solar rays. We do not want the space which is not reached by the solar rays, though it be in itself of an enormous extent. That which is not reached by the rays is not reached by the perception of the senses, and that which is not reached by perception is not knowable,

may be based ultimately upon this passage.

The reading *bham* of our text must be incorrect. It is a reading adopted by Parameśvara who was determined to prove that Āryabhaṭa did not teach the rotation of the Earth. This passage could not be explained away by recourse to false knowledge (*mithyājñāna*) as could I, 1 and IV, 9 and therefore was changed. The true reading is *bhūḥ*, as is proved conclusively by the quotation of Brahmagupta (XI, 17):

> prāṇenaiti kalāṁ bhūr yadi tarhi kuto vrajet kam adhvānam |
> āvarttanam urvyāś cen na patanti samucchrayāḥ kasmāt. ||

Compare Brahmagupta (XXI, 59) and Alberuni (I, 276–77, 280).

5. A *yojana* consists of 8,000 times a *nṛ* [the height of a man]. The diameter of the Earth is 1,050 *yojanas*. The diameter of the Sun is 4,410 *yojanas*. The diameter of the Moon is 315 *yojanas*. Meru is one *yojana*. The diameters of Venus, Jupiter, Mercury, Saturn, and Mars are one-fifth, one-tenth, one-fifteenth, one-twentieth, and one-twenty-fifth of the diameter of the Moon. The years of a *yuga* are equal to the number of revolutions of the Sun in a *yuga*.

As pointed out by Bhāu Dājī,[1] Brahmagupta (XI, 15–16) seems to quote from this stanza in his criticism of the diameter of the Earth given by Āryabhaṭa

ṣoḍaśagaviyojana paridhiṁ pratibhūvyāsaṁ pulāvadatā |
ātmajñānaṁ khyāpitam aniścayas tanikṛtakanyāt ||
bhūvyāsasyājñānād vyarthaṁ deśāntaraṁ tadajñānāt |
sphuṭatithyantājñānaṁ tithināśād grahaṇayor nāśaḥ. ||

The text of Brahmagupta is corrupt and must be emended. See the commentary of Sudhākara, who suggests for the first stanza

nṛṣiyojanabhūparidhiṁ prati bhūvyāsaṁ punar ñilā vadatā |
ātmajñānaṁ khyāpitam aniścayas tatkṛtavyāsaḥ. ||

Lalla (*Madhyamādhikāra*, 56 and *Candragrahaṇādhikāra*, 6) gives the same diameters for the Earth and the Sun but gives 320 as the diameter of the Moon, and (*Grahayutyadhikāra*, 2) gives for the planets the same fractions of the diameter of the Moon.[2]

Alberuni (I, 168) quotes from Brahmagupta Āryabhaṭa's diameter of the Earth, and a confused

[1] *JRAS*, 1865, p. 402.

[2] Cf. *Sūryasiddhānta*, I, 59; IV, 1; VII, 13–14; Brahmagupta, XXI, 32; Kharegat (*op. cit.*, XIX, 132–34, discussing *Sūryasiddhānta*, IX, 15–16).

passage (I, 244–46) quotes Balabhadra on Āryabhaṭa's conception of Meru. Its height is said to be a *yojana*. The context of the foregoing stanza seems to imply that its diameter is a *yojana*, as Parameśvara takes it. It is probable that its height is to be taken as the same.

If Parameśvara is correct in interpreting *samārkasamāḥ* as *yugasamā yugārkabhagaṇasamā*, the nominative plural *samāḥ* has been contracted after *sandhi*.

6. The greatest declination of the ecliptic is 24 degrees. The greatest deviation of the Moon from the ecliptic is 4½ degrees, of Saturn 2 degrees, of Jupiter 1 degree, of Mars 1½ degrees, of Mercury and Venus 2 degrees. Ninety-six *aṅgulas* or 4 *hastas* make 1 *nṛ*.

Parameśvara explains the words *bhāpakramo grahāṁśāḥ* as follows: *grahāṇāṁ bha aṁśāś caturviṁśatibhāgā apakramaḥ. paramāpakrama ity arthaḥ.* The construction is as strange as that of stanza 4 above.[1]

7. The ascending nodes of Mercury, Venus, Mars, Jupiter, and Saturn having moved (are situated) at 20, 60, 40, 80, and 100 degrees from the beginning of Meṣa. The apsides of the Sun and of the above-mentioned planets (in the same order) (are situated) at 78, 210, 90, 118, 180, and 236 degrees from the beginning of Meṣa.

I have followed Parameśvara's explanation of *gatvāṁśakān* as *uktān etān evāṁśakān meṣādito gatvā vyavasthitāḥ*.

In view of IV, 2, "the Sun and the nodes of the planets and of the Moon move constantly along the

[1] Cf. *Sūryasiddhānta*, I, 68–70 and II, 28; Brahmagupta, IX, 1 and XXI, 52.

ecliptic," and of I, 2, which gives the number of revolutions of the node of the Moon in a *yuga*, the word *gatvā* ("having gone") seems to imply, as Parameśvara says, a knowledge of the revolution of the nodes of the planets and to indicate that Āryabhaṭa intended merely to give their positions at the time his treatise was composed. The force of *gatvā* continues into the second line and indicates a knowledge of the revolutions of the apsides.

Āryabhaṭa gives figures for the revolutions of the apsis and node of the Moon. Other *siddhāntas* give figures for the revolutions of the nodes and apsides of all the planets. These seem to be based on theory rather than on observation since their motion (except in the case of the Moon) is so slow that it would take several thousand years for them to move so far that their motion could easily be detected by ordinary methods of observation.[1] Āryabhaṭa may have refrained from giving figures for the revolutions of nodes and apsides (except in the case of the Moon) because he distrusted the figures given in earlier books as based on theory rather than upon accurate observation. Brahmagupta XI, 8 (quoted above to stanza 2) remarks in criticism of Āryabhaṭa that in the *Daśagītika* the nodes are stationary while in the *Āryāṣṭaśata* they move. This refers to I, 2 and IV, 2. In the *Daśagītika* only the revolutions of the nodes of the Moon are given; in the *Āryāṣṭaśata* the nodes and apsides are said explicitly to move along the ecliptic. In the present stanza the word *gatvā* seems clearly to

[1] Cf. *Sūryasiddhānta*, pp. 27–28.

indicate a knowledge of the motion of the nodes and apsides of the other planets too. If Āryabhaṭa had intended to say merely that the nodes and apsides are situated at such-and-such places the word *gatvā* is superfluous. In a text of such studied brevity every word is used with a very definite purpose. It is true that Āryabhaṭa regarded the movement of the nodes and apsides of the other planets as negligible for purposes of calculation, but Brahmagupta's criticism seems to be captious and unjustified (see also Brahmagupta, XI, 6–7, and the commentary of Sudhākara to XI, 8). Barth's criticism[1] is too severe.

Lalla (*Spaṣṭādhikāra*, 9 and 28) gives the same positions for the apsides of the Sun and five planets (see also *Pañcasiddhāntikā*, XVII, 2).

For the revolutions of the nodes and apsides see Brahmagupta, I, 19–21, and *Sūryasiddhānta*, I, 41–44, and note to I, 44.

8. Divided by $4\frac{1}{2}$ the epicycles of the apsides of the Moon, the Sun, Mercury, Venus, Mars, Jupiter, and Saturn (in the first and third quadrants) are 7, 3, 7, 4, 14, 7, 9; the epicycles of the conjunctions of Saturn, Jupiter, Mars, Venus, and Mercury (in the first and third quadrants) are 9, 16, 53, 59, 31;

9. the epicycles of the apsides of the planets Mercury, Venus, Mars, Jupiter, and Saturn in the second and fourth quadrants are 5, 2, 18, 8, 13; the epicycles of the conjunctions of the planets Saturn, Jupiter, Mars, Venus, and Mercury in the second and fourth quadrants are 8, 15, 51, 57, 29. The circumference within which the Earth-wind blows is 3,375 *yojanas*.

The criticism of these stanzas made by Brahmagupta (II, 33 and XI, 18–21) is, as pointed out by

[1] *Op. cit.*, III, 154.

Sudhākara, not justifiable. For the dimensions of Brahmagupta's epicycles see II, 34–39).

Lalla (*Spaṣṭādhikāra*, 28) agrees closely with stanza 8 and (*Grahabhramaṇa*, 2) gives the same figure for the Earth-wind. Compare also *Sūryasiddhānta*, II, 34–37 and note, and *Pañcasiddhāntikā*, XVII, 1, 3.

10. The (twenty-four) sines reckoned in minutes of arc are 225, 224, 222, 219, 215, 210, 205, 199, 191, 183, 174, 164, 154, 143, 131, 119, 106, 93, 79, 65, 51, 37, 22, 7.

In Indian mathematics the "half-chord" takes the place of our "sine." The sines are given in minutes (of which the radius contains 3,438) at intervals of 225 minutes. The numbers given here are in reality not the values of the sines themselves but the differences between the sines.

Compare Sūryasiddhānta (II, 15–27) and Lalla (*Spaṣṭādhikāra*, 1–8) and Brahmagupta (II, 2–9). Bhāskara (*Gaṇitādhyāya, Spaṣṭādhikāra, Vāsanābhāṣya* to 3–9) refers to the *Sūryasiddhānta* and to Āryabhaṭa as furnishing a precedent for the use of twenty-four sines.[1]

Krishnaswami Ayyangar[2] furnishes a plausible explanation of the discrepancy between certain of the values given in the foregoing stanza and the values as calculated by II, 12.[3] Some of the discrepancies may be due to bad readings of the manuscripts. Kern

[1] For discussion of the stanza see Barth, *ibid.*, III, 150 n., and *JRAS*, 1911, pp. 123–24.

[2] *JIMS*, XV (1923–24), 121–26.

[3] See also Naraharayya, "Note on the Hindu Table of Sines," *ibid.*, pp. 105–13 of "Notes and Questions."

in a footnote to the stanza and Ayyangar (p. 125 n.) point out that the text-reading for the sixteenth and seventeenth sines violates the meter. This, however, may be remedied easily without changing the values.[1]

C. Whoever knows this *Daśagītika Sūtra* which describes the movements of the Earth and the planets in the sphere of the asterisms passes through the paths of the planets and asterisms and goes to the higher Brahman.

[1] Cf. *JRAS*, 1910, pp. 752, 754, and *IA*, XX, 228.

CHAPTER II

GAṆITAPĀDA OR MATHEMATICS

1. Having paid reverence to Brahman, the Earth, the Moon, Mercury, Venus, the Sun, Mars, Jupiter, Saturn, and the asterisms, Āryabhaṭa sets forth here [in this work] the science which is honored at Kusumapura.[1]

The translation "here at Kusumapura the revered science" is possible. At any rate, Āryabhaṭa states the school to which he belongs. Kusumapura may or may not have been the place of his birth.

2. The numbers *eka* [one], *daśa* [ten], *śata* [hundred], *sahasra* [thousand], *ayuta* [ten thousand], *niyuta* [hundred thousand], *prayuta* [million], *koṭi* [ten million], *arbuda* [hundred million], and *vṛnda* [thousand million] are from place to place each ten times the preceding.[2]

The names for classes of numbers are given only to ten places, although I, B describes a notation which reaches at least to the eighteenth place. The highest number actually used by Āryabhaṭa himself runs to ten places.

3. A square, the area of a square, and the product of two equal quantities are called *varga*. The product of three equal quantities, and a solid which has twelve edges are called *ghana*.[3]

[1] Translated by Fleet, *JRAS*, 1911, p. 110. See Kern's Preface to his edition of the *Bṛhat Saṁhitā*, p. 57, and *BCMS*, XVIII (1927), 7.

[2] See *JRAS*, 1911, p. 116; *IHQ*, III, 112; *BCMS*, XVII (1926), 198. For the quotation in Alberuni (I, 176), which differs in the last two names, see the criticism in *BCMS*, XVII (1926), 71.

[3] Read *dvādaśāśras* with Parameśvara. For *aśra* in the sense of "edge" see Colebrooke, *Algebra*, pp. 2 n. and 280 n. The translations given by Rodet and Kaye are inaccurate.

4. One should always divide the *avarga* by twice the (square) root of the (preceding) *varga*. After subtracting the square (of the quotient) from the *varga* the quotient will be the square root to the next place.

Counting from right to left, the odd places are called *varga* and the even places are called *avarga*. According to Parameśvara, the nearest square root to the number in the last odd place on the left is set down in a place apart, and after this are set down the successive quotients of the division performed. The number subtracted is the square of that figure in the root represented by the quotient of the preceding division. The divisor is the square of that part of the root which has already been found. If the last subtraction leaves no remainder the square root is exact. "Always" indicates that if the divisor is larger than the number to be divided a zero is to be placed in the line (or a blank space left there). *Sthānāntare* ("in another place") is equivalent to the *paṅkti* ("line") of the later books.

This process seems to be substantially correct, but there are several difficulties. *Sthānāntare* may mean simply "to another place," that is to say, each division performed gives another figure of the root. *Nityam* ("always") may merely indicate that such is the regular way of performing the operation.

All the translators except Saradakanta Ganguly translate *vargād varge śuddhe* with what precedes. I think he is correct in taking it with what follows. In that case the parallelism with the following rule is exact. Otherwise the first rule would give the operation for the *varga* place and then that for the *avarga* place while the second rule would give first the operations for the *aghana* places and then that for the *ghana* place. However, for purposes of description, it makes no difference whether the operations are given in one or the other of these orders.

Parallelism with *ghanasya mūlavargeṇa* of the following rule seems to indicate that *vargamūlena* is not to be translated "square root" but "root of the (preceding) *varga*."

If the root is to contain more than two figures the *varga* of *vargamūlena* is to be interpreted as applying to all the preceding figures up to and including the *varga* place which is being worked with. That is to say, the word *mūla* would refer to the whole of that part of the root which had already been found.[1]

For discussion see Kaye,[2] Avadhesh Narayan Singh,[3] Saradakanta Ganguly.[4] I cannot agree with Ganguly's discussion of the words *bhāgaṁ hared avargāt*. I see no reason to question the use of *bhāgaṁ harati* with the ablative in the sense of "divide." Brahmagupta (XII, 7) in his description of the process of extracting the cube root has *chedo 'ghanād dvītīyāt,* which means "the divisor of the second aghana."

Kaye[5] insists that this rule and the next are perfectly general (i.e., algebraical) and apply to all arithmetical notations. He offers no proof and gives

[1] See Colebrooke, *op. cit.*, p. 280 n.

[2] *JASB,* 1907, pp. 493–94.

[3] *BCMS,* XVIII (1927), 124.

[4] *JBORS,* XII, 78.

[5] *Op. cit.*, 1908, p. 120.

no example of the working of the rule according to his interpretation. To what do the words "square" and "non-square" of his translation refer? The words of Āryabhaṭa exactly fit the method employed in later Indian mathematics. Although Brahmagupta does not give a rule for square root, his method for cube root is that described below, although the wording of his rule is different from that of Āryabhaṭa's. I fail to see any similarity to the rule and method of Theon of Alexandria.

In the following example the sign ° indicates the *varga* places, and the sign – indicates the *avarga* places.

	$\overset{\circ}{1}\overset{-}{5}\overset{\circ}{1}\overset{-}{2}\overset{\circ}{9}$ (root = 1
Square of the root	1
Twice the root (2×1)	2)05(2 = quotient (or next digit of root)
	4
	11
Square of the quotient	4
Twice the root (2×12)	24)72(3 = quotient (or next digit of root)
	72
	09
Square of the quotient	9
	0

Square root is 1 2 3

5. One should divide the second *aghana* by three times the square of the (cube) root of the (preceding) *ghana*. The square (of the quotient) multiplied by three times the *pūrva* (that part of the cube root already found) is to be subtracted from the first

aghana, and the cube (of the quotient of the above division) is to be subtracted from the *ghana*.

The translation given by Avadhesh Narayan Singh[1] as a "correct literal rendering" is inaccurate. There is nothing in the Sanskrit which corresponds to "after having subtracted the cube (of the quotient) from the *ghana* place" or to "the quotient placed at the next place gives the root." The latter thought, of course, does carry over into this rule from the preceding rule. In the same article (p. 132) the Sanskrit of the rule is inaccurately printed with *trighanasya* for *triguṇena ghanasya*.[2]

Kaye[3] remarks that this rule is given by Brahmagupta "word for word." As a matter of fact, the Sanskrit of the two rules is very different, although the content is exactly the same.

Counting from right to left, the first, fourth, etc., places are named *ghana* (cubic); the second, fifth, etc., places are called the first *aghana* (non-cubic) places; and the third, sixth, etc., places are called the second *aghana* (non-cubic) places. The nearest cube root to the number in (or up to and including) the last *ghana* place on the left is the first figure of the cube root. After it are placed the quotients of the successive divisions. If the last subtraction leaves no remainder the cube root is exact.

[1] *BCMS*, XVIII (1927), 134.

[2] The rule has been discussed in *JBORS*, XII, 80. Cf. Brahmagupta (XII, 7) and the translation and note of Colebrooke (*op. cit.*, p. 280).

[3] *Op. cit.*, 1908, p. 119.

In the following example the sign ° indicates the *ghana* places and the sign – indicates the *aghana* places.

	$\overset{\circ}{1}\overset{-}{8}\overset{-}{6}\overset{\circ}{0}\overset{-}{8}\overset{-}{6}\overset{\circ}{7}$ (root = 1
Cube of root	1
	—
Three times square of root (3×1²)	3)08(2 = quotient (or next digit of root)
	6
	—
	26
Square of quotient multiplied by three times the *pūrva* (2²×3×1)	12
	—
	140
Cube of quotient	8
	—
Three times square of root (3×12²)	432)1328(3 = quotient (or next digit of root)
	1296
	—
	326
Square of quotient multiplied by three times the *pūrva* (3²×3×12)	324
	—
	27
Cube of quotient	27
	—
	0

Cube root is 1 2 3

6. The area of a triangle is the product of the perpendicular and half the base. Half the product of this area and the height is the volume of a solid which has six edges (pyramid).

If *samadalakoṭī* can denote, as Parameśvara says, a perpendicular which is common to two triangles the rule refers to all triangles. If *samadalakoṭī* refers to a perpendicular which bisects the base it refers only to isosceles triangles.[1]

[1] For *aśra* or *aśri* in the sense of "edge" see note to stanza II, 3. See *JBORS*, XII, 84–85, for discussion of the inaccurate value given in the second part of the rule.

7. Half of the circumference multiplied by half the diameter is the area of a circle. This area multiplied by its own square root is the exact volume of a sphere.[1]

8. The two sides (separately) multiplied by the perpendicular and divided by their sum will give the perpendiculars (from the point where the two diagonals intersect) to the parallel sides.

The area is to be known by multiplying half the sum of the two sides by the perpendicular.

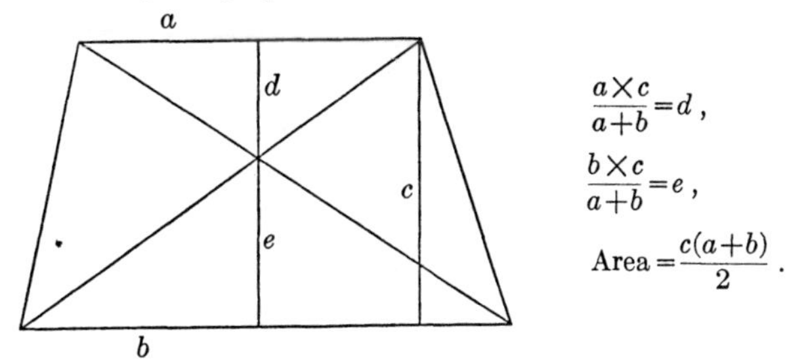

$$\frac{a \times c}{a+b} = d \,,$$

$$\frac{b \times c}{a+b} = e \,,$$

$$\text{Area} = \frac{c(a+b)}{2} \,.$$

The rule applies to any four-sided plane figure of which two sides are parallel, i.e., trapezium. The word translated "sides" refers to the two parallel sides. The perpendicular is the perpendicular between the two parallel sides.

In the example given above a and b are the parallel sides, c is the perpendicular between them, and d and e are the perpendiculars from the point of intersection of the two diagonals to the sides a and b, respectively.

9. The area of any plane figure is found by determining two sides and then multiplying them together.

The chord of the sixth part of the circumference is equal to the radius.

[1] See *ibid.* and *Bibl. math.*, IX, 196, for discussion of the inaccurate value given in the second part of the rule. For a possible reference to this passage by Bhāskara, *Golādhyāya*, *Bhuvanakośa*, stanza 61 (*Vāsanābhāṣya*) (not stanza 52 as stated), see *BCMS*, XVIII (1927), 10.

The very general rule given in the first half of this stanza seems to mean, as Parameśvara explains in some detail, that the mathematician is to use his ingenuity in determining two sides which will represent the average length and the average breadth of the figure. Their product will be the area. Methods to be employed with various kinds of figures were doubtless handed down by oral tradition.

Rodet thinks that the rule directs that the figure be broken up into a number of trapeziums. It is doubtful whether the words can bear that interpretation.

10. Add 4 to 100, multiply by 8, and add 62,000. The result is approximately the circumference of a circle of which the diameter is 20,000.

The circumference is 62,832. The diameter is 20,000.

By this rule the relation of circumference to diameter is 3.1416.[1]

Bhāskara, *Golādhyāya, Bhuvanakośa* (stanza 52), *Vāsanābhāṣya,* refers to this rule of Āryabhaṭa.

11. One should divide a quarter of the circumference of a circle (into as many equal parts as are desired). From the triangles and quadrilaterals (which are formed) one will have on the radius as many sines of equal arcs as are desired.[2]

The exact method of working out the table is not known. It is uncertain what is intended by the triangle and the quadrilateral constructed from each point marked on the quadrant.[3]

[1] See *JBORS*, XII, 82; *JRAS*, 1910, pp. 752, 754.

[2] See the table given in I, 10 of the differences between the sines. Twenty-four sines taken at intervals of 225 minutes of arc are regularly given in the Indian tables.

[3] Note the methods suggested by Kaye and Rodet and cf. *JIMS*, XV (1923–24), 122 and 108–9 of "Notes and Questions."

12. By what number the second sine is less than the first sine, and by the quotient obtained by dividing the sum of the preceding sines by the first sine, by the sum of these two quantities the following sines are less than the first sine.

The last phrase may be translated "the sine-differences are less than the first sine."[1]

This rule describes how the table of sine-differences given in I, 10 may be calculated from the first one (225). The first sine means always this first sine 225. The second sine means any particular sine with which one is working in order to calculate the following sine.

Subtract 225 from 225 and the remainder is 0. Divide 225 by 225 and the quotient is 1. The sum of 0 and 1 is subtracted from 225 to obtain the second sine 224. Subtract 224 from 225 and the remainder is 1. Divide 225 plus 224 by 225 and the nearest quotient is 2. Add 2 and 1 and subtract from 225. The third sine will be 222. Proceed in like manner for the following sines.

If this method is followed strictly there results several slight divergences from the values given in I, 10. It is possible to reconcile most of these by assuming, as Krishnaswami Ayyangar does, that from time to time the neglected fractions were distributed among the sines. But of this there is no indication in the rule as given.

[1] For discussion of the Indian sines see the notes of Rodet and Kaye; *Pañcasiddhāntikā*, chap. iv; *Sūryasiddhānta*, II, 15–27; Lalla, p. 12; Brahmagupta, II, 2–10; *JRAS*, 1910, pp. 752, 754; *IA*, XX, 228; Brennand, *Hindu Astronomy*, pp. 210–13; *JIMS*, XV (1923–24), 121–26, with attempted explanation of the variation of several of the values given in the table from the values calculated by means of this rule, and *ibid.*, pp. 105–13 of "Notes and Questions."

How Kaye gets "If the first and second be bisected in succession the sine of the half-chord is obtained" is a puzzle to me. It is impossible as a translation of the Sanskrit.

13. The circle is made by turning, and the triangle and the quadrilateral by means of a *karṇa;* the horizontal is determined by water, and the perpendicular by the plumb-line.

Tribhuja denotes triangle in general and *caturbhuja* denotes quadrilateral in general. The word *karṇa* regularly denotes the hypotenuse of a right-angle triangle and the diagonal of a square or rectangle. I am not sure whether the restricted sense of *karṇa* limits *tribhuja* and *caturbhuja* to the right-angle triangle and to the square and rectangle or whether the general sense of *tribhuja* and *caturbhuja* generalizes the meaning of *karṇa* to that of one chosen side of a triangle and to that of the diagonal of any quadrilateral. At any rate, the context shows that the rule deals with the actual construction of plane figures.

Parameśvara interprets it as referring to the construction of a triangle of which the three sides are known and of a quadrilateral of which the four sides and one diagonal are known. One side of the triangle is taken as the *karṇa.* Two sticks of the length of the other two sides, one touching one end and the other the other end of the *karṇa,* are brought to such a position that their tips join. The quadrilateral is made by constructing two triangles, one on each side of the diagonal.

The circle is made by the turning of the *karkaṭa* or compass.[1]

14. Add the square of the height of the gnomon to the square of its shadow. The square root of this sum is the radius of the *khavṛtta*.

The text reads *khavṛtta* ("sky-circle"). Parameśvara reads *svavṛtta* ("its circle"). I do not know which is correct.

Kaye remarks that in order "to mark out the hour angles on an ordinary sun-dial, it is necessary to describe two circles, one of which has its radius equal to the vertical gnomon and the other with radius equal to the hypotenuse of the triangle formed by the equinoctial shadow and the gnomon." It may be that this second circle is the one referred to here. Parameśvara has *chāyāgramadhyaṁ śankuśiraḥprāpi yan maṇḍalam ūrdhvādhaḥsthitam tat svavṛttam ity ucyate*, "the circle which has its centre at the extremity of the shadow and which touches the top of the gnomon is called the *svavṛtta*." As Rodet remarks, it is difficult to see for what purpose such a circle could serve.

15. Multiply the length of the gnomon by the distance between the gnomon and the *bhujā* and divide by the difference between the length of the gnomon and the length of the *bhujā*. The quotient will be the length of the shadow measured from the base of the gnomon.[2]

[1] For parallels to the stanza see Lalla (*Yantrādhyāya*, 2) and Brahmagupta, XXII, 7. See *BCMS*, XVIII (1927), 68–69, which is too emphatic in its assertion that *karṇa* must mean "diagonal" and not "hypotenuse."

[2] See Brahmagupta, XII, 53; Colebrooke, *op. cit.*, p. 317; Brennand, *op. cit.*, p. 166.

Because of the use of the word *koṭī* in the following rule Rodet is inclined to think that the gnomon and the *bhujā* were not perpendicular but projected horizontally from a wall. *Bhujā* denotes any side of a triangle, but *koṭī* usually refers to an upright. It is possible, however, for *koṭī* to denote any perpendicular to the *bhujā* whether horizontal or upright.

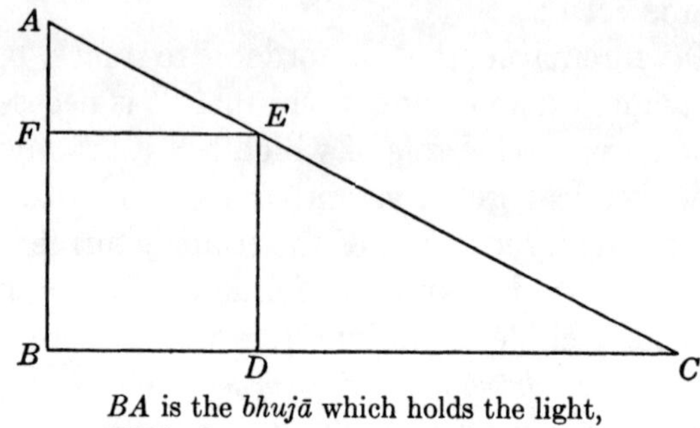

BA is the *bhujā* which holds the light,
DE is the gnomon,

$$DC=\frac{DE\times BD}{AF}.$$

16. The distance between the ends of the two shadows multiplied by the length of the shadow and divided by the difference in length of the two shadows gives the *koṭī*. The *koṭī* multiplied by the length of the gnomon and divided by the length of the shadow gives the length of the *bhujā*.

The literal translation of *chāyāguṇitaṁ chāyāgravivaram ūnena bhājitā koṭī* seems to be "The distance between the ends of the two shadows multiplied by the length of the shadow is equal to the *koṭī* divided by the difference in length of the two shadows." This is equivalent to the translation given above.

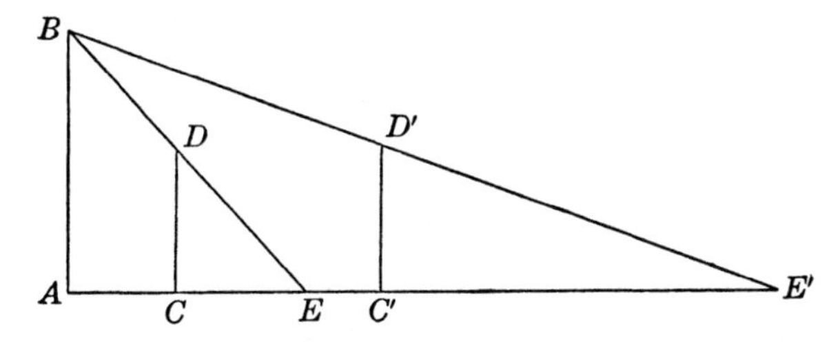

AB is the *bhujā*,
AE is the *koṭī*,
CD is the gnomon in its first position,
C'D' is the gnomon in its second position,
CE and *C'E'* are the first and second shadows,

$$AE = \frac{CE \times EE'}{C'E' - CE},$$

$$AB = \frac{AE \times CD}{CE}.$$

The length of the *bhujā* which holds the light and the distance between the end of the shadow and the base of the *bhujā* are unknown. In order to find them the gnomon is placed in another position so as to give a second shadow.

The length of the shadow is its length when the gnomon is in its first position. The *koṭī* is the distance between the end of the shadow when the gnomon is in its first position and the base of the *bhujā*.

The word *koṭī* means perpendicular (or upright) and the rule might be interpreted, as Rodet takes it, as meaning that the *bhujā* and the gnomon extend horizontally from a perpendicular wall. But the words *bhujā* and *koṭī* also refer to the sides of a right-angle triangle without much regard as to which is horizontal and which is upright.

Or the first position of the gnomon may be $C'D'$ and the second CD. To find AE' and AB.[1]

17. The square of the *bhujā* plus the square of the *koṭī* is the square of the *karṇa*.

In a circle the product of two *śaras* is the square of the half-chord of the two arcs.

The *bhujā* and *koṭī* are the sides of a right-angle triangle. The *karṇa* is the hypotenuse.

The *śaras* or "arrows" are the segments of a diameter which bisects any chord.[2]

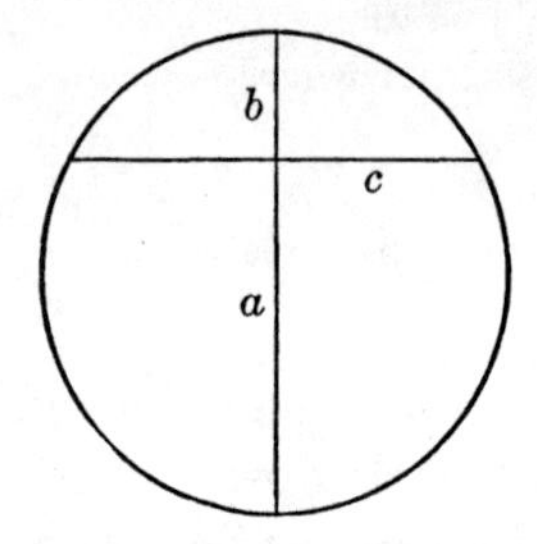

$a \times b = c^2$
where c is the half-chord.

18. (The diameters of) two circles (separately) minus the *grāsa*, multiplied by the *grāsa*, and divided separately by the sum of (the diameters of) the two circles after the *grāsa* has been subtracted from each, will give respectively the *sampātaśaras* of the two circles.

When two circles intersect the word *grāsa* ("the bite") denotes that part of the common diameter of the two circles which is cut off by the intersecting chords of the two circles.

[1] See Brahmagupta, XII, 54; Colebrooke, *op. cit.*, p. 318; Brennand, *op. cit.*, p. 166.

[2] Cf. Brahmagupta, XII, 41. See *BCMS*, XVIII (1927), 11, 71, with discussion of the quotation given by Colebrooke, *op. cit.*, p. 309, from Pṛthūdakasvāmī's commentary to Brahmagupta.

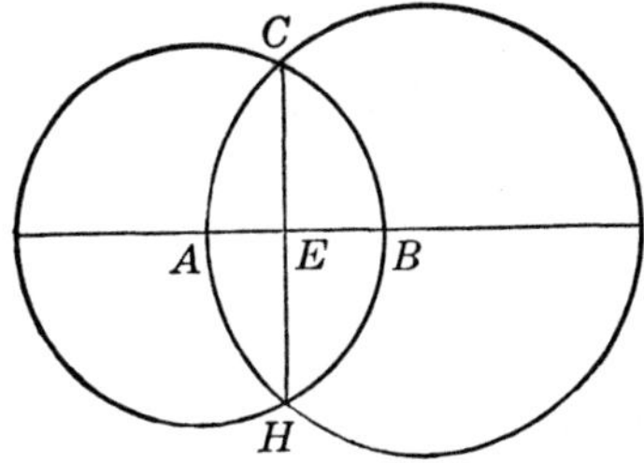

AB is the *grāsa*,
AE and BE are the
sampātaśaras.

$$AE=\frac{AB(d-AB)}{D+d-2AB}, \qquad EB=\frac{AB(D-AB)}{D+d-2AB},$$

where d and D are the diameters of the two circles.

The *sampātaśaras* are the two distances (within the *grāsa*), on the common diameter, from the circumferences of the two circles to the point of intersection of this common diameter with the chord connecting the two points where the circumferences intersect.[1]

19. The desired number of terms minus one, halved, plus the number of terms which precedes, multiplied by the common difference between the terms, plus the first term, is the middle term. This multiplied by the number of terms desired is the sum of the desired number of terms.

Or the sum of the first and last terms is multiplied by half the number of terms.

This rule tells how to find the sum of any desired number of terms taken anywhere within an arithmetical progression. Let n be the number of terms extending from the $(p+1)$th to the $(p+n)$th terms in an arithmetical progression, let d be the common difference between the terms, let a be the first term of the progression, and l the last term.

[1] Cf. Brahmagupta, XII, 43; Colebrooke, *op. cit.*, p. 311.

The second part of the rule applies only to the sum of the whole progression beginning with the first term.

$$S=n\left[a+\left(\frac{n-1}{2}+p\right)d\right],$$

$$S=\frac{(a+l)n}{2}.$$

As Parameśvara says, *samukhamadhyam* must be taken as equivalent to *samukhaṁ madhyam.*

Whether Parameśvara is correct in his statement *bahusūtrārthapradarśakam etat sūtram. ato bahudhā yojanā kāryā* and subsequent exposition seems very doubtful.

Brahmagupta, XII, 17 has only the second part of the rule.[1]

20. Multiply the sum of the progression by eight times the common difference, add the square of the difference between twice the first term and the common difference, take the square root of this, subtract twice the first term, divide by the common difference, add one, divide by two. The result will be the number of terms.

$$n=\frac{1}{2}\left[\frac{\sqrt{8dS+(d-2a)^2}-2a}{d}+1\right].$$ [2]

As Rodet says, the development of this formula from the one in the preceding rule seems to indicate knowledge of the solution of quadratic equations in the form $ax^2+bx+c=0$.[3]

[1] Cf. Colebrooke, *op. cit.*, p. 290.

[2] See Brahmagupta, XII, 18; Colebrooke, *op. cit.*, p. 291.

[3] See also *JA* (1878), I, 28, 77, and *JBORS*, XII, 86–87.

21. In the case of an *upaciti* which has one for the first term and one for the common difference between the terms the product of three terms having the number of terms for the first term and one as the common difference, divided by six, is the *citighana*. Or the cube of the number of terms plus one, minus the cube root of this cube, divided by six.

Form an arithmetical progression 1 2 3 4 5, etc. Form the series 1 3 6 10 15, etc., by taking for the terms the sum of the terms of the first series. The rule gives the sum of this series.

It also gives the cubic contents of a pile of balls which has a triangular base. The wording of the rule would seem to imply that it was intended especially for this second case. *Citighana* means "cubic contents of the pile," and *upaciti* ("pile") refers to the base (or one side) of the pile, i.e., 1 2 3 4 5, etc.[1]

As Rodet remarks, it is curious that in the face of this rule the rule given above (stanza 6) for the volume of a pyramid is incorrect.

$$S=\frac{n(n+1)(n+2)}{6} \quad \text{or} \quad \frac{(n+1)^3-(n+1)}{6}.$$

22. The sixth part of the product of three quantities consisting of the number of terms, the number of terms plus one, and twice the number of terms plus one is the sum of the squares. The square of the sum of the (original) series is the sum of the cubes.

From the series 1 2 3 4, etc., form the series 1 4 9 16, etc., and 1 8 27 64, etc., consisting of the

[1] Cf. Brahmagupta, XII, 19; Colebrooke, *op. cit.*, pp. 292–93. Brahmagupta (XII, 20) directs that the summation of certain series be illustrated by means of piles of round balls.

squares and cubes of the terms of the first series. The rule tells how to find the sums of the second and third series.[1]

The rule for finding the sum of the first series was given above in stanza 19.

The sum of the squares is

$$\frac{n(n+1)(2n+1)}{6}.$$

23. One should subtract the sum of the squares of two factors from the square of their sum. Half the result is the product of the two factors.

$$ab=\frac{(a+b)^2-(a^2+b^2)}{2}.$$

24. Multiply the product (of two factors) by the square of two (4), add the square of the difference between the two factors, take the square root, add and subtract the difference between the two factors, and divide the result by two. The results will be the two factors.

$$\frac{\sqrt{4ab+(a-b)^2}\pm(a-b)}{2}\text{ will give } a \text{ and } b.$$

25. Multiply the sum of the interest on the principal and the interest on this interest by the time and by the principal. Add to this result the square of half the principal. Take the square root of this. Subtract half the principal and divide the remainder by the time. The result will be the interest on the principal.[2]

[1] Cf. Brahmagupta, XII, 20; Colebrooke, *op. cit.*, p. 293; *BCMS*, XVIII (1927), 70.

[2] Cf. the somewhat similar problem in Brahmagupta, XII, 15; Colebrooke, *op. cit.*, pp. 287–28, and see the discussion of Kaye.

The formula involves the solution of a quadratic equation in the form of $ax^2+2bx=c$.[1]

A sum of money is loaned. After a certain unit of time the interest received is loaned for a known number of units of time at the same interest. What is known is the amount of the interest on the principal plus the interest on this interest. Call this B. Let the principal be A. Let t be the time.

$$x=\frac{\sqrt{B\times A\times t+\left(\frac{A}{2}\right)^2}-\frac{A}{2}}{t}.$$

The following example is given by Parameśvara. The sum of 100 is loaned for one month. Then the interest received is loaned for six months. At that time the original interest plus the interest on this interest amounts to 16.

$$\begin{array}{l} B=16 \\ A=100 \\ t=6 \end{array} \quad \frac{\sqrt{16\times100\times6+2500}-50}{6}=10.$$

The interest received on 100 in one month was 10.

26. In the rule of three multiply the fruit by the desire and divide by the measure. The result will be the fruit of the desire.

The rule of three corresponds to proportion.

In the proportion a is to b as c is to x the measure is a, the fruit is b, the desire is c, and the fruit of the desire is x.

$$x=\frac{bc}{a}.$$ [2]

[1] See *JBORS*, XII, 87.

[2] See Brahmagupta, XII, 10 and Colebrooke, *op. cit.*, p. 283.

27. The denominators of multipliers and divisors are multiplied together. Multiply numerators and denominators by the other denominators in order to reduce fractions to a common denominator.

For the first part of the rule I have given what seems to be the most likely literal translation. The exact sense is uncertain. Kaye (agreeing with Rodet) translates, "The denominators are multiplied by one another in multiplication and division." If that is the correct translation the genitive plural is curious. Parameśvara explains *guṇakāra* as *guṇaguṇyayor āhatir atra guṇakāraśabdena vivakṣitā. hārya ity arthaḥ* and then seems to take *bhāgahāra* as referring to a fractional divisor of this product. Can the words bear that construction? In either case the inversion of numerator and denominator of the divisor would be taken for granted.

It is tempting to take *guṇakārabhāgahāra* as meaning "fraction" and to translate, "The denominators of fractions are multiplied together." But for that interpretation I can find no authority.

28. Multipliers become divisors and divisors become multipliers, addition becomes subtraction and subtraction becomes addition in the inverse method.

The inverse method consists in beginning at the end and working backward. As, for instance, in the question, "What number multiplied by 3, divided by 5, plus 6, minus 1, will give 5?"

29. If you know the results obtained by subtracting successively from a sum of quantities each one of these quantities set these results down separately. Add them all together and divide by the number of terms less one. The result will be the sum of all the quantities.

The translation given by Kaye is incorrect. The revised translation given in his *Indian Mathematics*, page 47, is not an improvement.[1]

$x-d=a+b+c$
$x-a=b+c+d$
$x-b=a+c+d$
$x-c=a+b+d$

According to the rule $3a+3b+3c+3d$ divided by 3 gives $x=a+b+c+d$ since $4x=4a+4b+4c+4d$.

30. One should divide the difference between the pieces of money possessed by two men by the difference between the objects possessed by them. The quotient will be the value of one of the objects if the wealth of the two men is equal.

Two men possess each a certain number of pieces of money (such as rupees) and a certain number of objects of merchandise (such as cows).

Let a and b be the number of rupees possessed by two men, and let m and p be the number of cows possessed by them.

$$x=\frac{b-a}{m-p} \text{ since } mx+a=px+b\,.$$

If one man has 100 rupees and 6 cows and the other man has 60 rupees and 8 cows the value of a cow is 20 rupees provided the wealth of the two men is equal.

31. The two distances between two planets moving in opposite directions is divided by the sum of their daily motions. The two distances between two planets moving in the same direction is divided by the difference of their daily motions. The two results (in each case) will give the time of meeting of the two in the past and in the future.

[1] Cf. *JBORS*, XII, 88–90.

In each case there will be two distances between the planets, namely, that between the one which is behind and the one which is ahead, and, measuring in the same direction, the distance between the one which is ahead and the one which is behind. This seems to be the only adequate interpretation of the word *dve*. The translations of Rodet and Kaye fail to do full justice to the word *dve*.[1]

The next two stanzas give a method for the solution of indeterminate equations of the first degree; but no help for the interpretation of the process intended, which is only sketchily presented in Āryabhaṭa, is to be found in Mahāvīra, Bhāskara, or the second Āryabhaṭa. The closest parallel is found in Brahmagupta, XVIII, 3–5.[2] The verbal expression is very similar to that of Āryabhaṭa, but with one important exception. In place of the enigmatic statement *matiguṇam agrāntare kṣiptam*, "(The last remainder) is multiplied by an assumed number and added to the difference between the *agras*," Brahmagupta has, "The residue (of the reciprocal division) is multiplied by an assumed number such that the product having added to it the difference of the remainders may be exactly divisible (by the residue's divisor). That multiplier is to be set down (underneath) and the quotient last." It is possible that this same process is to be understood in Āryabhaṭa.

[1] Cf. Parameśvara, *dve iti vacanam antarasya dvaividhyāt. śīghragatihīno mandagatir antaraṁ bhavati. mandagatihīnaś śīghragatiś cāntaraṁ bhavati. iti dvaividhyam* and his further interpretation of the results.

Cf. Brahmagupta, IX, 5–6 and Bhāskara, *Gaṇitādhyāya*, *Grahayutyadhikāra*, 3–4, and *Vāsanābhāṣya*, and see *JA* (1878), I, 28.

[2] Colebrooke, *op. cit.*, p. 325.

First I shall explain the stanza on the basis of Parameśvara's interpretation and of Brahmagupta's method:

32–33. Divide the divisor which gives the greater *agra* by the divisor which gives the smaller *agra*. The remainder is reciprocally divided (that is to say, the remainder becomes the divisor of the original divisor, and the remainder of this second division becomes the divisor of the second divisor, etc.). (The quotients are placed below each other in the so-called chain.) (The last remainder) is multiplied by an assumed number and added to the difference between the *agras*. Multiply the penultimate number by the number above it and add the number which is below it. (Continue this process to the top of the chain.) Divide (the top number) by the divisor which gives the smaller *agra*. Multiply the remainder by the divisor which gives the greater *agra*. Add this product to the greater *agra*. The result is the number which will satisfy both divisors and both *agras*.

In this the sentence, "(The last remainder) is multiplied by an assumed number and added to the difference between the *agras*," is to be understood as equivalent to the quotation from Brahmagupta given above.

The word *agra* denotes the remainders which constitute the provisional values of x, that is to say, values one of which will satisfy one condition, one of which will satisfy the second condition of the problem. The word *dvicchedāgra* denotes the value of x which will satisfy both conditions.

I cannot agree with the translation given by Kaye (and followed by Mazumdar, *BCMS*, III, 11) or accept the method given by Kaye. Kaye's translation of *matiguṇam agrāntare kṣiptam*, "An assumed number together with the original difference is thrown in," is an impossible translation, and any method based on that translation is bound to be incorrect. It omits altogether the important word *guṇam* ("multiplied"). Since the preceding phrase dealt with the remainders of the reciprocal division, the natural word to supply with *matiguṇam* seems to be *śeṣam* ("remainder"). Something has to be supplied, and Brahmagupta's method offers a possible interpretation. A second possible interpretation, which will be given below, would supply "quotient" instead of "remainder."

The following example is given by Parameśvara.

$\frac{8x}{29}$ gives a remainder of 4 $\qquad$ $\frac{17x}{45}$ gives a remainder of 7.

These are equivalent to $\frac{8x-4}{29}=y$ and $\frac{17x-7}{45}=z$ or $8x-29y=4$ and $17x-45z=7$
where y and z are the quotients of the division (y and z to be whole numbers).

1. First process. To find a value of x which will satisfy the first equation:

```
8)29(3
  24
  --
   5)8(1
     5
     -
     3)5(1
       3
       -
       2)3(1
         2
         -
         1
```

Take an assumed number such that multiplied by 1 (the last remainder of the reciprocal division) and plus or minus 4 (the original remainder) it will be exactly divisible by 2 (the last divisor of the reciprocal division).

$$6 \text{ is taken because } \frac{6-4}{2}=1.$$

Therefore 6 and 1 are to be added to the quotients to form the chain.

```
3  73                29)73(2
1  20                   58
1  13                   --
1   7                   15
6
1
```

This remainder 15 is the *agra*, that is to say, a value of x which will satisfy the equation.

2. Second process. To find a value of x which will satisfy the second equation:

```
17)45(2
   34
   --
   11)17(1
      11
      --
       6)11(1
          6
         --
          5)6(1
            5
            -
            1
```

Take an assumed number such that multiplied by 1 (the last remainder of the reciprocal division) and plus or minus 7 (the original remainder) it will be exactly divisible by 5 (the last divisor of the reciprocal division).

$$3 \text{ is taken because } \frac{3+7}{5}=2.$$

Therefore 3 and 2 are to be added to the quotients to form the chain.

```
2  34                              34)45(1
1  13                                 34
1   8                                 —
1   5                                 11
3     This remainder 11 is the agra, that is to say, a value of x
2        which will satisfy the equation.
```

These numbers 15 and 11 are the *agras* mentioned at the beginning of the rule. The corresponding divisors are 29 and 45. The difference between the *agras* is 4, i.e., $15-11$.

3. Third process. To find a value of x which will satisfy both equations:

```
29)45(1
   29
   —
   16)29(1
      16
      —
      13)16(1
         13
         —
          3)13(4
            12
            —
             1
```

Take an assumed number such that multiplied by 1 (the last remainder of the reciprocal division) and plus or minus 4 (the difference between the agras) it will be exactly divisible by 3 (the last divisor of the reciprocal division).

2 is taken because $\frac{2+4}{3}=2$.

Therefore 2 and 2 are to be added to the quotients to form the chain.

1	34	45)34(0
1	22	0
1	12	—
4	10	34
2		Therefore 34 is the remainder.
2		

Then in accordance with the rule $34\times 29=986$ and $986+15=1001$

This number 1001 is the smallest number which will satisfy both equations.

Strictly speaking, the rule applies only to the third process given above. The solution of the single indeterminate equation is taken for granted and is not given in full. There is nothing to indicate how far the reciprocal division was to be carried. Must it be carried to the point where the last remainder is 1? Must the number of quotients taken to make the chain be even in number?

On page 50 of Kern's edition a 1 has been omitted by mistake (twice) as the fourth member of the chains given.

The following method was partially worked out by Mazumdar,[1] who was misled in some details by following Kaye's translation, and by Sen Gupta,[2] and fully worked out by Sarada Kanta Ganguly.[3]

[1] *BCMS*, III, 11–19.

[2] *Journal of the Department of Letters* (Calcutta University), XVI, 27–30.

[3] *BCMS*, XIX (1928), 170–76.

According to Ganguly's interpretation, the translation would be:

32–33. Divide the divisor corresponding to the greater remainder by the divisor corresponding to the smaller remainder. The remainder (and the divisor) are reciprocally divided. (This process is continued until the last remainder is 0.) (The quotients are placed below each other in the so-called chain.) Multiply any assumed number by the last quotient of the reciprocal division and add it to the difference between the two remainders. (Interpreted as meaning that this product and this difference are placed in the chain beneath the quotients.) Multiply the penultimate number by the number above it and add the number which is below it. (Continue this process to the top of the chain.) Divide (the lower of the two top numbers) by the divisor corresponding to the smaller remainder. Multiply the remainder by the divisor corresponding to the greater remainder. Add the product to the greater remainder. The result is the (least number) which will satisfy the two divisors and the two remainders.

He remarks:

The implication is that the least number satisfying the given conditions can also be obtained by multiplying the remainder, obtained as the result of division of the upper number by the divisor corresponding to the greater given remainder, by the divisor corresponding to the smaller given remainder and then adding the smaller remainder to the product.

From this point of view the problem would be that of finding a number which will leave given remainders when divided by given positive integers.

For example, to take a simple case: What number divided by 3 and 7 will leave as remainders 2 and 1?

Following the rule literally, even though a smaller

number has to be divided by a larger number we get the following:

```
7)3(0
  0
  —
  3)7(2
    6
    —
    1)3(3
      3
      —
      0
```

Multiply the last quotient (3) by an assumed number (for instance, 3) and set this product and the difference between the remainders 2 and 1, i.e., (1) down below the quotients to form the chain.

```
0  28                      7)65(9
2  65                        63
3  28                        —
9                             2
1
```

Then $2 \times 3 = 6$ and $6 + 2 = 8$.

or

```
3)28(9
  27
  —
   1
```

Then $1 \times 7 = 7$ and $7 + 1 = 8$.

Therefore 8 is the number desired.

The two methods attach different significations to the word *agra* and supply different words with *bhājite* in the third line ("remainder," in one case; "quotient," in the other). They differ fundamentally in their interpretations of the words *matiguṇam agrā-*

ntare kṣiptam. In the first method it is necessary to supply much to fill out the meaning, but the translation of these words themselves is a more natural one. In the second method it is not necessary to supply anything except "quotient" with *matiguṇam* (in the first method it is necessary to supply "remainder"). But if the intention was that of stating that the product of the quotient and an assumed number, *and* the difference between the remainders, are to be added below the quotients to form a chain the thought is expressed in a very curious way. Ganguly finds justification for this interpretation (p. 172) in his formulas, but I cannot help feeling that the Sanskrit is stretched in order to make it fit the formula.

The general method of solution by reciprocal division and formation of a chain is clear, but some of the details are uncertain and we do not know to what sort of problems Āryabhaṭa applied it.

CHAPTER III

KĀLAKRIYĀ OR THE RECKONING OF TIME

1. A year consists of twelve months. A month consists of thirty days. A day consists of sixty *nāḍīs*. A *nāḍī* consists of sixty *vināḍikās*.[1]

2. Sixty long letters or six *prāṇas* make a sidereal *vināḍikā*. This is the division of time. In like manner the division of space beginning with a revolution.[2]

3. The difference between the number of revolutions of two planets in a *yuga* is the number of their conjunctions.

Twice the sum of the revolutions of the Sun and Moon is the number of *vyatīpātas*.[3]

This is a *yoga* of the Sun and Moon when they are in different *ayanas*, have the same declination, and the sum of their longitudes is 180 degrees.

4. The difference between the number of revolutions of a planet and the number of revolutions of its *ucca* is the number of revolutions of its epicycle.

The number of revolutions of Jupiter multiplied by 12 are the years of Jupiter beginning with Aśvayuja.[4]

[1] Cf. *Sūryasiddhānta*, I, 11–13; Alberuni, I, 335; Bhaṭṭotpala, p. 24.

[2] Cf. *Sūryasiddhānta*, I, 11, 28; Bhaṭṭotpala, p. 24; *Pañcasiddhāntikā*, XIV, 32, for the first part of 2; Brahmagupta, I, 5–6, and Bhāskara, *Gaṇitādhāya*, *Kālamānādhyāya*, 16–18, for both stanzas.

[3] See Lalla, *Madhyamādhikāra*, 11; Brahmagupta, XIII, 42, for the first part. For *vyatīpāta* see *Sūryasiddhānta*, XI, 2; *Pañcasiddhāntikā*, III, 22; Lalla, *Mahāpātādhikāra*, 1; Brahmagupta, XIV, 37, 39.

[4] For the first part see Lalla, *Madhyamādhikāra*, 11; Brahmagupta, XIII, 42; Bhāskara, *Gaṇitādhyāya*, *Bhagaṇādhyāya*, 14. For the second part see *JRAS*, 1863, p. 378; *ibid.*, 1865, p. 404; *Sūryasiddhānta*, I, 55; Bhaṭṭotpala, p. 182.

The word *ucca* refers both to *mandocca* ("apsis") and *śīghrocca* ("conjunction").

Parameśvara explains that the number of revolutions of the epicycle of the apsis of the Moon is equal to the difference between the number of revolutions of the Moon and the revolutions of its apsis; that since the apsides of the six others are stationary, the number of revolutions of the epicycles of their apsides is equal to the number of revolutions of the planets; and that the number of revolutions of the epicycles of the conjunctions of Mercury, Venus, Mars, Jupiter, and Saturn is equal to the difference between the revolutions of the planets and the revolutions of their conjunctions.

As pointed out in the note to I, 7, the apsides were not regarded by Āryabhaṭa as being stationary in the absolute sense. They were regarded by him as stationary for purposes of calculation at the time when his treatise was composed since their movements were very slow.

5. The revolutions of the Sun are solar years. The conjunctions of the Sun and Moon are lunar months. The conjunctions of the Sun and the Earth are [civil] days. The revolutions of the asterisms are sidereal days.

The word *yoga* applied to the Sun and the Earth (instead of *bhagaṇa* or *āvarta*) seems clearly to indicate that Āryabhaṭa believed in a rotation of the Earth (see IV, 48). Parameśvara's explanation, *ravibhūyogaśabdena raver bhūparibhramaṇam abhihitam*, seems to be impossible.

6. Subtract the solar months in a *yuga* from the lunar months in a *yuga*. The result will be the number of intercalary months in

a *yuga*. Subtract the natural [civil] days in a *yuga* from the lunar days in a *yuga*. The result will be the number of omitted lunar days in a *yuga*.[1]

7. A solar year is a year of men. Thirty of these make a year of the Fathers. Twelve years of the Fathers make a year of the gods.

8. Twelve thousand years of the gods make a *yuga* of all the planets. A thousand and eight *yugas* of the planets make a day of Brahman.[2]

9. The first half of a yuga is called *utsarpiṇī* [ascending]. The latter half is called *avasarpiṇī* [descending]. The middle part of a yuga is called *suṣamā*. The beginning and the end are called *duṣṣamā*. Because of the apsis of the Moon.

Alberuni (I, 370–71) remarks:

Āryabhaṭa of Kusumapura, who belongs to the school of the elder Āryabhaṭa, says in a small book of his on *Al-ntf* (?), that "1,008 *caturyugas* are one day of Brahman. The first half of 504 *caturyugas* is called *utsarpiṇī*, during which the sun is ascending, and the second half is called *avasarpiṇī*, during which the sun is descending. The midst of this period is called *sama*, i.e., equality, for it is the midst of the day, and the two ends are called *durtama* (?).

"This is so far correct, as the comparison between day and *kalpa* goes, but the remark about the sun's ascending and descending is not correct. If he meant the sun who makes our day, it was his duty to explain of what kind that ascending and descending of the sun is; but if he meant a sun who specially belongs to the day of Brahman, it was his duty to show or to describe him to us. I almost think that the author meant by these two expressions the progressive, increasing development of things during the first half of this period, and the retrograde, decreasing development in the second half."

[1] Cf. *Sūryasiddhānta*, I, 35–36; Lalla, *Madhyamādhikāra*, 10; Brahmagupta, I, 24 and XIII, 26.

[2] Cf. *Sūryasiddhānta*, I, 13–15; I, 20. Brahmagupta (I, 12) criticizes Āryabhaṭa's figure of 1,008 *yugas* instead of 1,000 *yugas*. Cf. *JRAS*, 1865, p. 400. Cf. also I, 3 and see *JRAS*, 1911, p. 486.

The reference is to the foregoing stanza. The middle of the *yuga* seems to be called *suṣamā* ("even") because good and bad are evenly mixed. The beginning and the end are called *duṣṣamā* ("uneven") because in one case goodness and in the other case badness predominates.

Parameśvara remarks that the *vyākhyākāra* has given no explanation. Then he quotes from the *Bhaṭaprakāśikā* a statement to the effect that our text refers to the increase and decrease of men's lives in the course of a *yuga* and a criticism (*asyārtho 'bhiyuktair nirūpya vaktavyaḥ*) of the last phrase of the stanza. He then continues by saying that he does not see what meaning can be intended by the word *indūccāt*, and adds that the word has nothing to do with the matter under discussion, has no significance for the calculation of the places of the planets. Then he adds two forced explanations. The meaning of *indūccāt* is quite uncertain.

Sudhākara (*Gaṇakataraṅgiṇī*, p. 7) suggests the emendation to *agnyaṁśāt*.

The terminology is distinctively Jaina.[1]

10. When three *yugapādas* and sixty times sixty years had elapsed (from the beginning of the *yuga*) then twenty-three years of my life had passed.[2]

If Āryabhaṭa began the Kaliyuga at 3102 B.C. as later astronomers did, and if his fourth *yugapāda*

[1] See Hemacandra, *Abhidhānacintāmaṇi*, 128–35; Glasenapp, *Der Jainismus*, pp. 244–45; Kirfel, *Kosmographie*, p. 339; *ZDMG*, LX, 320–21; Stevenson, *The Heart of Jainism*, pp. 272 ff. See also Hardy, *Manual of Buddhism*, p. 7.

[2] See *JRAS*, 1863, p. 387; *ibid.*, 1865, p. 405; Kern, *Bṛhat Saṁhitā*, Preface, p. 57; *JRAS*, 1911, pp. 111–12.

began with the beginning of the Kaliyuga, we arrive at the date 499 A.D. It is natural to take this as the date of composition of the treatise. Parameśvara quotes the *Prakāśikākāra* to the effect that this is to be taken as the date at which the calculations of the true places of the planets made by it would be correct, and that for later times a correction would have to be made.

The word *iha* may mean "here" or "now." Parameśvara takes it as referring to this present twenty-eighth *caturyuga*.

11. The *yuga*, the year, the month, and the day began all together at the beginning of the bright fortnight of Caitra. Time, which has no beginning and no end, is measured by (the movements of) the planets and the asterisms on the sphere.

Bhāu Dājī[1] first pointed out the criticism made of this stanza by Brahmagupta (XI, 6):

yugavarṣādīn vadatā caitrasitādeḥ samaṁ pravṛttān yat|
tad asad yataḥ sphuṭayugaṁ tat sthairyān mandapātānām.||

Compare Brahmagupta, I, 4, and Bhāskara, *Gaṇitādhyāya*, *Kālamānādhyāya*, 15, who refers to an earlier commentary in which time is called endless.[2]

12. The planets moving equally (traversing the same distance in *yojanas* each day) in their orbits complete the circle of the asterisms in sixty solar years, and the circle of the sky in a divine age [*caturyuga*].

In sixty years they move a distance in *yojanas* equal to the circle of the asterisms. In a *caturyuga* they move a distance in *yojanas* equal to the circumference of the sky (*ākāśakakṣyā*) (cf. I, 4).

[1] *JRAS*, 1865, p. 401.

[2] For discussion of the stanza see Fleet, *ibid.*, 1911, pp. 489–90; cf. I, 2.

The planets really all move at the same speed. The nearer ones seem to move more rapidly than the more distant ones because their orbits are smaller.[1]

13. The Moon, being below, completes its small orbit in a short time. Saturn, being above all the others, completes its large orbit in a long time.[2]

14. The zodiacal signs (a twelfth of the circle) are to be known as small in a small circle and large in a large circle. Likewise the degrees and minutes are the same in number in the various orbits.[3]

15. Beneath the asterisms are Saturn, Jupiter, Mars, the Sun, Venus, Mercury, and the Moon, and beneath these is the Earth situated in the center of space like a hitching-post.[4]

16. These seven lords of the hours, Saturn and the others, are in order swifter than the preceding one, and counting successively the fourth in the order of their swiftness they become the Lords of the days from sunrise.

They are called "swifter than the preceding" because their orbits being successively smaller they complete their revolutions in less time (traverse a given number of degrees in less time). The order of the planets is Saturn, Jupiter, Mars, the Sun, Venus, Mercury, and the Moon. Therefore they become rulers of the days of the week as follows:

[1] Cf. *Sūryasiddhānta*, I, 27 and note; Brahmagupta, XXI, 12; *Pañcasiddhāntikā*, XIII, 39; Bhāskara, *Golādhyāya*, *Bhuvanakośa*, 69; *JRAS*, 1911, p. 112.

[2] Cf. *JRAS*, 1863, p. 375; *Sūryasiddhānta*, XII, 76–77; *Pañcasiddhāntikā*, XIII, 41; Brahmagupta, XXI, 14; Bhaṭṭotpala, p. 45.

[3] Cf. *JRAS*, 1863, p. 375; *Pañcasiddhāntikā*, XIII, 40; *Sūryasiddhānta*, XII, 75; Brahmagupta, XXI, 14; Bhaṭṭotpala, p. 45.

[4] Cf. *JRAS*, 1863, p. 375; *Pañcasiddhāntikā*, XIII, 39; Lalla, *Madhyamādhikāra*, 12; Brahmagupta, XXI, 2; Bhaṭṭotpala, p. 44.

Saturday—Saturn
Sunday—Sun
Monday—Moon
Tuesday—Mars
Wednesday—Mercury
Thursday—Jupiter
Friday—Venus

For the first part see Brahmagupta, XXI, 13; *Sūryasiddhānta*, XII, 78.[1]

Bhāu Dājī[2] first pointed out the criticism of this stanza made by Brahmagupta (XI, 12):

sūryādayaś caturthā dinavāra yad uvāca tad asad āryabhaṭaḥ|
laṅkodaye yato 'rkasyāstamayaṁ prāha siddhapure. ||

As Sudhākara shows, the criticism is a futile one.

17. All the planets move by their (mean) motion on their orbits and their eccentric circles from the apsis eastward and from the conjunction westward.[3]

The mean planet moves with its mean motion on its orbit the center of which is the center of the Earth. The true planet moves with its (mean) motion on an eccentric circle the center of which does not coincide with the center of the Earth.

Kakṣyā in this passage stands for *kakṣyāmaṇḍala*, the orbit on which the mean planet moves. The *pratimaṇḍala* is the eccentric circle on which the true planet moves. Because of the eccentricity of this second circle the planet is sometimes seen ahead of and sometimes back of its mean place.[4]

[1] See Barth (*Œuvres*, III, 151) concerning this as the only reference to astrology in Āryabhaṭa's treatise. The reference to *vyatīpāta* (III, 5) should be added.

[2] *JRAS*, 1865, p. 401.

[3] See Lalla, *Chedyakādhikāra*, 12–13; Brahmagupta, XIV, 11 and XXI, 24.

[4] See Brennand, *Hindu Astronomy*, pp. 224 ff.; *Sūryasiddhānta*, p. 64.

18. The eccentric circle of each planet is equal to its *kakṣyā-maṇḍala* [the orbit on which the mean planet moves]. The center of the eccentric circle is outside the center of the solid Earth.

The *kakṣyāmaṇḍala* is determined by I, 4.

19. The distance between the center of the Earth and the center of the eccentric circle is equal to the radius of the epicycle. The planets move with their mean motions on their epicycles.[1]

Brahmagupta, XI, 52, has

nīcoccavṛttamadhyasya golabāhyena nāma kṛtam uccam |
tatstho na bhavati ucco yatas tato vetti noccam api. ||

If this really refers to Āryabhaṭa the criticism is futile since Āryabhaṭa does not call the center of the epicycle *ucca*. As Caturvedācārya says, *vāgbalam etat.*

20. The planet in its swift motion from its *ucca* has a *pratiloma* motion on its epicycle. In its slow motion from its *ucca* it has an *anuloma* motion on its epicycle.

The exact meaning of this is not clear to me. It can hardly mean that the planet moves on its epicycle *pratiloma* from its *śīghrocca* and *anuloma* from its *mandocca*. On the epicycle of the apsis the motion should be exactly the reverse of these.[2]

Anuloma means "eastward" or "ahead"; *Pratiloma* means "westward" or "behind."

Parameśvara remarks that *anuloma* and *pratiloma* refer to the planet's position with reference to the mean planet as ahead of it or behind it. He also remarks that the planet is *śīghragati* in the six signs

[1] Cf. Lalla, *Chedyakādhikāra*, 8–9; Brahmagupta, XIV, 10 and XXI, 24–26.

[2] See Brahmagupta, XXI, 25–26 and *Sūryasiddhānta*, pp. 63–64, 67–68.

which are above, and *mandagati* in the six signs which are below the *ucca*. When *pratiloma* the true planet is below the mean planet. When *anuloma* the true planet is above the mean planet.[1]

madhyamakakṣāvṛtte madhyamayā gacchati graho gatyā |
upariṣṭhāt tallaghvyā tadadhikagatyā tv adhaḥsthaḥ syāt. ||

Parameśvara sums up the content of the stanza with *madhyamāt sphuṭasya pratilomānulomagatitvam uktam.*

The meaning of the stanza seems to be that during half of its revolution on its epicycle the planet is ahead of the mean planet and during half of its revolution is behind the mean planet.

21. The epicycles move eastward from the apsis and westward from the conjunctions. The mean planet, situated on its orbit, appears at the center of its epicycle.[2]

The next three stanzas state briefly the method of calculating the true places of the planets from their mean places.

Parameśvara explains the method as follows:

For the Sun and Moon only one process of correction is required, that for the apsis.

For Mars, Jupiter, and Saturn four processes are necessary: (1) From the mean place the *mandaphala* is calculated and (half of it is) applied to the mean place. (2) From this corrected place the *śīghraphala* is calculated and half of it is applied to the corrected place. (3) From this result the *mandaphala* is again calculated and applied to the mean place. (4) From this result the *śīghraphala* is again calculated and applied to the place obtained in the third process.

[1] Cf. Lalla, *Bhuvanakośa*, 38. [2] Cf. Brahmagupta, XXI, 25.

For Venus and Mercury three processes are necessary: (1) From the mean place the *śīghraphala* is calculated and half of it is applied *vyastam* (in reverse order) to the *mandocca* (apsis). (2) This corrected *mandocca* is subtracted from the mean place, the *mandaphala* is calculated from this and applied to the mean place. (3) From this corrected place the *śīghraphala* is calculated and applied to the place obtained in the second process.

22–23. (The corrections) from the apsis are minus, plus, plus, minus (in the four quadrants). (The corrections) from the conjunctions are just the reverse.

In the case of Saturn, Jupiter, and Mars in the first process half of the *mandaphala* obtained from the apsis is minus and plus to the mean planet. Half (the correction) from the conjunction is minus and plus to the *manda* planets. (By applying the correction) from the apsis they become *sphuṭamadhya.* (By applying the correction) from the conjunction they become *sphuṭa.*

24. Half (the correction) from the conjunction is to be applied minus and plus to the apsis. (By applying the correction) from the *manda* [apsis] thus obtained Venus and Mercury become *sphuṭamadhya.* They become *sphuṭa* (by applying the correction from the conjunction).

The first half of stanza 22 gives the general rule as to whether the equations of anomaly and of commutation (*mandaphala* and *śīghraphala*) are to be added or subtracted in each of the four quadrants. The equation from the apsis is minus in the half of the orbit beginning with Meṣa, plus in the half of the orbit beginning with Tula. The equation from the conjunction is plus in the half of the orbit beginning with Mesa, minus in the half of the orbit beginning with Tula.

The planet is called *manda* after the first correction from the apsis has been applied to the mean place. *Sphuṭa* means "true."

In stanza 24 Parameśvara gives no explanation of the two last words, *sphuṭau bhavataḥ*. It would be natural to take these words as summing up what precedes and to understand that only two processes are involved. But Parameśvara's detailed description of the process in his commentary to stanza 21 indicates that three processes are involved, that *sphuṭau bhavataḥ* indicates a further application of the equation from the conjunction. The commentary to stanza 24 gives in detail the process of calculating the equations for apsis and conjunction.[1]

Brahmagupta (II, 19, 33, 46–47) criticizes Āryabhaṭa for the inaccuracy of his method of calculating the true places.

25. The product of its hypotenuses divided by the radius will give the distance between the planet and the Earth.

The planet has the same speed on its epicycle that it has on its orbit.

Parameśvara explains that the *karṇas* referred to are the *śīghrakarṇa* and the *mandakarṇa* employed in the last and the next to the last processes for calculating the true places of the planets.

The second half of the stanza is uncertain. This same statement was made in unmistakable terms in

[1] See *Pañcasiddhāntikā*, XVII, 4–10; *Sūryasiddhanta*, II, 43–45; Brahmagupta, II, 34–40; Lalla, *Spaṣṭādhikāra*, 31–36; Bhāskara, *Gaṇitādhyāya*, *Spaṣṭādhikāra*, 34–36 and *Golādhyāya*, *Chedyakādhikāra*, 10 ff.; *JRAS*, 1863, pp. 353–59; Brennand, *op. cit.*, pp. 214–28; Kaye, *Hindu Astronomy*, pp. 87–89.

III, 19. Parameśvara quotes the author of the earlier Prakāśikā, *bhūtārāgrahavivaravyāsārdhaviracitāyāṁ kakṣyāyāṁ yo grahasya javas sa mandanīcocce bhavati. tāvatpramāṇāyāṁ kakṣyāyāṁ graho mandasphuṭagatyā gacchatīty arthaḥ. ity āha. asmān kiṁ tv etan nopapannam iti pratibhāti.* Then he explains that the meaning may be that the radius of the epicycle is equal to the greatest distance by which the mean orbit lies inside or outside of the eccentric circle.

Grahavegaḥ is reminiscent of *grahajavaḥ* in I, 4, but the meaning can hardly be the same.

Karṇa ("hypotenuse") is the distance between the center of the Earth and the planet.[1]

[1] Cf. Brahmagupta, XXI, 31; Bhāskara, *Gaṇitādhyāya, Candragrahaṇādhikāra,* 4–5; *Sūryasiddhānta,* p. 69.

CHAPTER IV

GOLA OR THE SPHERE

1. From the beginning of Meṣa to the end of Kanyā is the northern half of the ecliptic. The other half from the beginning of Taulya to the end of Mīna is the southern half of the ecliptic. Both deviate equally from the Equator.

Therefore the greatest declinations north and south are equal, and the declinations of the first three signs in each half are equal to the declinations of the last three signs taken in reverse order.[1]

2. The Sun, the nodes of the planets, and the node of the Moon move constantly along the ecliptic. The shadow of the Earth moves along the ecliptic at a distance of 180 degrees from the Sun.

Bhāu Dājī[2] first pointed out the reference to this passage made by Brahmagupta, XI, 8.[3]

Barth[4] questions the stanza, but without good reason.

3. The Moon, from its nodes, moves northward and southward of the ecliptic. Likewise Jupiter, Mars, and Saturn. Venus and Mercury do the same from their conjunctions.[5]

4. When the Moon has no declination it is visible when 12 degrees from the Sun. Venus when 9 degrees. The other planets

[1] Cf. *JRAS*, 1863, p. 374; Bhaṭṭotpala, p. 45.

[2] *JRAS*, 1865, p. 401.

[3] Cf. I, 7 and note; Brahmagupta, XXI, 53; *Sūryasiddhānta*, IV, 6.

[4] *Œuvres*, III, 154.

[5] Cf. *Sūryasiddhānta*, I, 68–69 and *Āryabhaṭīya*, I, 6.

in succession according to their decreasing sizes when at 9 degrees increased by two's.

Compare Brahmagupta, VI, 6; *Sūryasiddhānta*, IX, 6–9 and X, 1; *Pañcasiddhāntikā*, XVII, 12 and XVIII, 58. Bhāu Dājī[1] first pointed out the criticism of this stanza made by Brahmagupta, VI, 12:

āryabhaṭaḥ kṣetrāṁśair dṛśyādṛśyān yad uktavāṁs tad asat |
dṛggaṇitavisaṁvādād dṛggaṇitaikyaṁ svakālāṁśaiḥ. ||

5. Half of the spheres of the Earth, the planets, and the asterisms is darkened by their shadows, and half, being turned toward the Sun, is light (being small or large) according to their size.[2]

6. The sphere of the Earth, being quite round, situated in the center of space, in the middle of the circle of asterisms, surrounded by the orbits of the planets, consists of water, earth, fire, and air.[3]

7. Just as a ball formed by a Kadamba flower is surrounded on all sides by blossoms just so the Earth is surrounded on all sides by all creatures terrestrial and aquatic.[4]

8. During a day of Brahman the sphere of the Earth increases a *yojana* in size all around. During a night of Brahman, which is equal in length to a day of Brahman, there is a decrease by the same amount of the Earth which has been increased by Earth.[5]

9. As a man in a boat going forward sees a stationary object moving backward just so at Laṅkā a man sees the stationary asterisms moving backward (westward) in a straight line.

The natural interpretation of this stanza seems

[1] *JRAS*, 1865, p. 401.

[2] Cf. Lalla, *Madhyagativāsanā*, 40–41; Bhaṭṭotpala, p. 100; *Pañcasiddhāntikā*, XIII, 35, for the Moon.

[3] Cf. III, 15. Cf. Lalla, *Bhūgolādhyāya*, 1; *Pañcasiddhāntikā*, XIII, 1; Bhaṭṭotpala, p. 58 (and see *JRAS*, 1863, pp. 373–74); Alberuni, I, 268.

[4] Cf. Lalla, *Bhūgolādhyāya*, 6; Bhaṭṭotpala, p. 58 (and see *JRAS*, 1863, pp. 373–74); Bhāskara, *Golādhyāya, Bhuvanakośa*, 3.

[5] Cf. Lalla, *Grahabhramasaṁsthādhyāya*, 20; Bhāskara, *Golādhyāya, Bhuvanakośa*, 62.

to be that an observer at the Equator of the Earth, which rotates toward the East, sees the stationary celestial objects as though moving westward. But Parameśvara explains that whereas the Earth does not really move, it appears to move toward the east because of the westward movement of the asterisms. He is forced to take the words *anuloma* and *viloma*, which regularly mean "ahead," "eastward," and "backward," "westward," in exactly the opposite senses. He explains that persons on the asterisms, which move toward the west, would seem to see stationary objects on the Earth moving eastward. As Barth[1] points out, this explanation is quite unacceptable. It seems that Parameśvara completely misrepresents the opinion of Āryabhaṭa, as clearly stated in several places in the text, and as described by Brahmagupta and other critics of Āryabhaṭa.

There is nothing to indicate that this stanza represents a state of affairs caused by *mithyājñāna* ("false knowledge)."

Bhaṭṭotpala (pp. 58–59) quotes this stanza and then refutes it by quoting the *Pañcasiddhāntikā*, XIII, 6–8, Pauliśa, Brahmagupta, and strangely enough Āryabhaṭa himself (the following stanza, IV, 10). It is curious that Āryabhaṭa should be quoted against himself, and that Bhaṭṭotpala should not indicate clearly which view really represents Āryabhaṭa's own opinion. It looks as though Bhaṭṭotpala regarded the first stanza as containing a *pūrvapakṣa* or erroneous view.[2]

[1] *Op. cit.*, III, 158 n.

[2] Cf. *JRAS*, 1863, pp. 375–77.

For criticisms of the rotation of the Earth see Alberuni (I, 276–77, 280); Lalla, *Mithyājñānādhyāya*, 42–43; Śrīpati as reported in the Lucknow edition of Bhāskara's *Golādhyāya*, page 83; see also Barth.[1]

Colebrooke[2] quotes Pṛthūdaka the commentator on Brahmagupta as follows:

> bhapañjaraḥ sthiro bhūr evāvṛtyāvṛtya prātidaivasikau |
> udayāstamayau sampādayati nakṣatragrahāṇām. ||[3]

The *Vāsanāvārttika* to Bhāskara's *Grahagaṇita*, page 113,[4] quotes the foregoing stanza and remarks that according to Āryabhaṭa the planets move toward the east, the asterisms are stationary, and the Earth rotates eastward.

10. The cause of their rising and setting is due to the fact that the circle of the asterisms, together with the planets, driven by the provector wind, constantly moves straight westward at Laṅkā.[5]

Bhaṭṭotpala (p. 59) quotes this stanza to disprove the preceding stanza which he quoted on page 58 (cf. *JRAS*, 1863, p. 377).[6]

The Marīci (p. 43) to Bhāskara's *Grahagaṇita*[7] quotes this stanza.

[1] *Op. cit.*, III, 158.

[2] *Essays*, II, 392.

[3] See *JRAS*, 1865, pp. 403–4; *IHQ*, I, 666 (the words given as a direct quotation from Āryabhaṭa are incorrect); *BCMS*, XVII (1926), 175. The author of the last article remarks that it is not clear whether Āryabhaṭa had in mind the geocentric or the heliocentric motion of the Earth. The latter is out of the question. Cf. III, 15, *bhūmir medhībhūtā khamadhyasthā*, and IV, 6, *khamadhyagataḥ*.

[4] *Pandit*, Vol. XXXI.

[5] Cf. *Sūryasiddhānta*, II, 3: Lalla, *Madhyamādhikāra*, 12.

[6] See Barth, *op. cit.*, III, 158.

[7] *Pandit*, Vol. XXX.

The *Vāsanāvārttika* to Bhāskara's *Grahagaṇita* (p. 118)[1] quotes this stanza apparently without seeing in it anything contradictory to the preceding stanza which was quoted on page 113, and with the remark that Āryabhaṭa is here following the opinion of Vṛddhavasiṣṭha.

If the readings of our text are correct it is difficult to see how the two stanzas can be brought into agreement. The ninth stanza states unequivocally that the asterisms are stationary and implies the rotation of the Earth. The tenth stanza seems to state that the asterisms, together with the planets, are driven by the provector wind. This would imply the ordinary point of view of most Indian astronomers that the Earth was stationary. Parameśvara avoids the difficulty by assuming that stanza 9 describes a state of mind brought about by *mithyājñāna* ("false knowledge"). But since several other stanzas (I, 1; I, 4; III, 5; IV, 48) and the testimony of later writers who quote Āryabhaṭa prove that Āryabhaṭa believed in the rotation of the Earth, it is impossible to follow Parameśvara. We might understand in stanza 10 the phrase "they seem to move" as stating a *pūrvapakṣa* (the erroneous view), but in the absence of any word to suggest this interpretation it is a doubtful expedient. Stanza 10 cannot be regarded as an interpolation (unless one stanza has been dropped out in order to make room for it) because the last three sections of Āryabhaṭa's work were known to Brahmagupta as "The Hundred and Eight Stanzas" (and our text contains 108 stanzas).

[1] *Ibid.*, Vol. XXXI.

11. In the center of the Nandana forest is Mount Meru, a *yojana* in measure (diameter and height), shining, surrounded by the Himavat Mountains, made of jewels, quite round.[1]

12. Heaven and Meru are at the center of the land, Hell and Vaḍavāmukha are at the center of the water. The gods and the dwellers in Hell both think constantly that the others are beneath them.[2]

Quoted by Bhaṭṭotpala, page 58.[3]

13. Sunrise at Laṅkā is sunset at Siddhapura, midday at Yavakoṭī, and midnight at Romaka.[4]

Brahmagupta's criticism (XI, 12)

sūryādayaś caturthā dinavārā yad uvāca tad asad āryabhaṭaḥ |
laṅkodaye yato 'rkasyāstamayaṁ prāha siddhapure ||

is incorrect, as pointed out by Sudhākara in his commentary.

14. Laṅkā is 90 degrees from the centers of the land and water [north and south poles]. Ujjain is straight north of Laṅkā by $22\frac{1}{2}$ degrees.[5]

15. From a level place half of the sphere of the asterisms

[1] Cf. I, 5. Cf. also *Sūryasiddhānta*, XII, 34; Lalla, *Bhuvanakośa*, 18–19; Bhāskara, *Golādhyāya, Bhuvanakośa*, 31; Alberuni (I, 244, 246). Quoted by Bhaṭṭotpala, p. 58 (cf. *JRAS*, 1863, p. 373).

[2] Cf. *Pañcasiddhāntikā*, XIII, 2–3; *Sūryasiddhānta*, XII, 35–36, 53; Brahmagupta, XXI, 3; Lalla, *Bhūgolādhyāya*, 3–4; Bhāskara, *Golādhyāya, Bhuvanakośa*, 17–20, 31.

[3] Cf. *JRAS*, 1863, p. 373.

[4] Cf. Kern, *Bṛhat Saṁhitā*, Preface, p. 57; *Sūryasiddhānta*, XII, 38–41; *Pañcasiddhāntikā*, XV, 23; Lalla, *Bhūgolādhyāya*, 12; Bhāskara, *Golādhyāya, Bhuvanakośa*, 17, 44; Alberuni I, 267–68; *JRAS*, 1865, p. 402.

[5] Cf. *Sūryasiddhānta*, I, 62; *Pañcasiddhāntikā*, XIII, 17; Lalla, *Madhyamādhikāra*, 55, and *Bhuvanakośa*, 41; Brahmagupta, XXI, 9; Bhāskara, *Golādhyāya, Bhuvanakośa*, 50 (*Vāsanābhāṣya*), and *Madhyagati*, 24; Alberuni, I, 316 (cf. *BCMS*, XVII [1926], 71).

decreased by the radius of the Earth is visible. The other half, plus the radius of the Earth, is cut off by the Earth.[1]

16. The gods, who dwell in the north on Meru, see the northern half of the sphere of the asterisms moving from left to right. The *Pretas*, who dwell in the south at Vaḍavāmukha, see the southern half of the sphere of the asterisms moving from right to left.[2]

Quoted by Bhaṭṭotpala, page 324.[3]

17. The gods and the *Pretas* see the Sun after it has risen for half a solar year. The Fathers who dwell in the Moon see it for half a lunar month. Here men see it for half a natural [civil] day.[4]

Referred to by Alberuni, I, 330.

18. There is a circle east and west (the prime vertical) and another north and south (the meridian) both passing through zenith and nadir. There is a horizontal circle, the horizon, on which the heavenly bodies rise and set.[5]

19. The circle which intersects the east and west points and two points on the meridian which are above and below the horizon by the amount of the observer's latitude is called the *unmaṇḍala*. On it the increase and decrease of day and night are measured.

The *unmaṇḍala* is the east and west hour-circle which passes through the poles. It is also called "the horizon of Laṅkā."[6]

[1] Cf. Lalla, *Bhuvanakośa*, 36; Brahmagupta, XXI, 64; Bhāskara, *Golādhyāya, Tripraśnavāsanā*, 38.

[2] Cf. *Sūryasiddhānta*, XII, 55; *Pañcasiddhāntikā*, XIII, 9; Brahmagupta, XXI, 6–7; Lalla, *Grahabhramasaṁsthādhyāya*, 3–5; Bhāskara, *Golādhyāya, Bhuvanakośa*, 51.

[3] Cf. *JRAS*, 1863, p. 378.

[4] Cf. *Sūryasiddhānta*, XII, 74 and XIV, 14; Lalla, *Grahabhramasaṁsthādyāya*, 14; Brahmagupta, XXI, 8; *Pañcasiddhāntikā*, XIII, 27, 38.

[5] Cf. Lalla, *Golabandhādhikāra*, 1–2; Brahmagupta, XXI, 49.

[6] Cf. Lalla, *Golabandhādhikāra*, 3; Brahmagupta, XXI, 50.

20. The east and west line and the north and south line and the perpendicular from zenith to nadir intersect in the place where the observer is.

21. The vertical circle which passes through the place where the observer is and the planet is the *dṛṅmaṇḍala*. There is also the *dṛkkṣepamaṇḍala* which passes through the nonagesimal point.[1]

The nonagesimal or central-ecliptic point is the point on the ecliptic which is 90 degrees from the point of the ecliptic which is on the horizon.

These two circles are used in calculating the parallax in longitude in eclipses.

22. A light wooden sphere should be made, round, and of equal weight in every part. By ingenuity one should cause it to revolve so as to keep pace with the progress of time by means of quicksilver, oil, or water.[2]

Sukumar Ranjan Das[3] remarks that two instruments are named in this stanza (the *gola* and the *cakra*). I can see no reference to the *cakra*.

23. On the visible half of the sphere one should depict half of the sphere of the asterisms by means of sines.

The equinoctial sine is the sine of latitude. The sine of colatitude is its *koṭi*.

The sine of the distance between the Sun and the zenith at midday of the equinoctial day is the equinoctial sine. This is the same as the equinoctial shadow and equals the sine of latitude. It is the base.

[1] Cf. *Sūryasiddhānta*, V, 6–7 n.; Kaye, *Hindu Astronomy*, p. 76.

[2] Cf. *Sūryasiddhānta*, XIII, 3 ff.; Lalla, *Yantrādhyāya*, 1 ff.; *IHQ*, IV, 265 ff.

[3] *IHQ*, IV, 259, 262.

The sine of co-latitude is the *koṭi* (the side perpendicular to the base) or *śaṅku* (gnomon).[1]

24. Subtract the square of the sine of the given declination from the square of the radius. The square root of the remainder will be the radius of the day-circle north or south of the Equator.

The day-circle is the diurnal circle of revolution described by a planet at any given declination from the Equator. So these day-circles are small circles parallel to the Equator.[2]

25. Multiply the day-radius of the circle of greatest declination (24 degrees) by the sine of the desired sign of the zodiac and divide by the radius of the day-circle of the desired sign of the zodiac. The result will be the equivalent in right ascension of the desired sign beginning with Meṣa.

To determine the right ascension of the signs of the zodiac, that is to say, the time which each sign of the ecliptic will take to rise above the horizon at the Equator.[3]

26. The sine of latitude multiplied by the sine of the given declination and divided by the sine of co-latitude is the earth-sine, which, being situated in the plane of one's day-circle, is the sine of the increase of day and night.

The earth-sine is the distance in the plane of the day-circle between the observer's horizon and the

[1] Cf. Brahmagupta, III, 7–8; Lalla, *Sāmānyagolabandha*, 9–10; Bhāskara, *Gaṇitādhyāya*, *Tripraśnādhikāra*, 12–13.

[2] Cf. Lalla, *Spaṣṭādhikāra*, 18; *Pañcasiddhāntikā*, IV, 23; *Sūryasiddhānta*, II, 60; Brahmagupta, II, 56; Bhāskara, *Gaṇitādhyāya*, *Spaṣṭādhikāra*, 48 (*Vāsanābhāṣya*); Kaye, *op. cit.*, p. 73.

[3] Cf. Lalla, *Tripraśnādhikāra*, 8; Brahmagupta, II, 57–58; *Sūryasiddhānta*, III, 42–43 and note; *Pañcasiddhāntikā*, IV, 29–30; Bhāskara, *Gaṇitādhyāya*, *Spaṣṭādhikāra*, 57; Kaye, *op. cit.*, pp. 79–80.

horizon of Laṅkā (the *unmaṇḍala*). When transformed to the plane of a great circle it becomes the ascensional difference.[1]

27. The first and fourth quadrants of the ecliptic rise in a quarter of a day (15 *ghaṭikās*) minus the ascensional difference. The second and third quadrants rise in a quarter of a day plus the ascensional difference, with regular increase and decrease.

The last phrase means that the values for signs 1, 2, 3 are equal, respectively, to those of signs 6, 5, 4 and that the values of 7, 8, 9 are equal, respectively, to those of 12, 11, 10. They increase in the first quadrant, decrease in the second, increase in the third, and decrease in the fourth. There are, therefore, only three numerical values involved, those calculated for the first three signs. See the table given in *Sūryasiddhānta*, III, 42–45 n.[2]

28. The sine of the Sun at any given point from the horizon on its day-circle multiplied by the sine of co-latitude and divided by the radius is the *śaṅku* when any given part of the day has elapsed or remains.

The *śaṅku* is the sine of the altitude of the Sun at any time on the vertical circle from the zenith passing through the Sun. Cf. Brahmagupta, XXI, 63, *dṛgmaṇḍale natāṁśajyā dṛgjyā śaṅkur unnatāṁśajyā*,

[1] Cf. *Sūryasiddhānta*, II, 61–63; Lalla, *Spaṣṭādhikāra*, 17, and *Sāmānyagolabandha*, 4; Brahmagupta, II, 57–60; *Pañcasiddhāntikā*, IV, 26 and note; Bhāskara, *Gaṇitādhyāya*, *Spaṣṭādhikāra*, 48; Kaye, *op. cit.*, p. 73.

[2] Cf. Lalla, *Madhyagativāsanā*, 15; Bhāskara, *Gaṇitādhyāya*, *Spaṣṭādhikāra*, 65 (*Vāsanābhāṣya*) who names Aryabhaṭa in connection with this rule.

and Bhāskara, *Golādhyāya*, *Tripraśnavāsanā*, 36, *śaṅkur unnatalavajyakā bhavet*.[1]

Parameśvara remarks: *uttaragole gatagantavyāsubhyaś caradalāsūn viśodhya jīvām ādāya svāhorātrārdhena nihatya trijyayā vibhajya labdhe bhūjyāṁ prakṣipet. sā kṣitijād utpannā svāhorātreṣṭajyā bhavati.* This corresponds to the so-called *cheda* of Brahmagupta.

29. Multiply the given sine of altitude of the Sun by the sine of latitude (the equinoctial sine) and divide by the sine of colatitude. The result will be the base of the *śaṅku* of the Sun south of the rising and setting line.

Śaṅkvagra is the same as *śaṅkutala* ("the base of the *śaṅku*") and denotes the distance of the base of the *śaṅku* from the rising and setting line.[2]

30. The sine of the greatest declination multiplied by the given base-sine of the Sun and divided by the sine of co-latitude is the Sun's *agrā* on the east and west horizons.

The *agrā* is the Sun's amplitude or the sine of the degrees of difference between the day-circle and the east and west points on the horizon.[3]

The proportions employed are those given in *Sūryasiddhānta*, V, 3 n.

[1] Cf. *Sūryasiddhānta*, III, 35–39 and note; Brahmagupta, III, 25–26; *BCMS*, XVIII (1927), 25.

[2] Cf. Brahmagupta, III, 65 and XXI, 63; Bhāskara, *Golādhyāya*, *Tripraśnavāsanā*, 40–42 (and *Vāsanābhāṣya*) and *Gaṇitādhyāya*, *Tripraśnādhikāra*, 73 (and *Vāsanābhāṣya*); Lalla, *Tripraśnādhikāra*, 49.

[3] See *Sūryasiddhānta*, III, 7 n.; Brahmagupta, XXI, 61; Bhāskara, *Golādhyāya*, *Tripraśnavāsanā*, 39 and *Gaṇitādhyāya*, *Tripraśnādhikāra*, 17 (*Vāsanābhāṣya*).

31. The measure of the Sun's amplitude north of the Equator [i.e., when the Sun is in the Northern hemisphere], if less than the sine of latitude, multiplied by the sine of co-latitude and divided by the sine of latitude gives the sine of the altitude of the Sun on the prime vertical.[1]

Bhāu Dājī[2] first pointed out that Brahmagupta (XI, 22) contains a criticism of stanzas 30–31.

uttaragole 'grāyāṁ viṣuvajjyāto yad uktam ūnāyām |
samamaṇḍalagas tad asat krāntijyāyāṁ yato bhavati ||

Parameśvara remarks: *viṣuvajjyonā cet. viṣuvajjyonayā krāntyā sādhitā ced ity arthaḥ. viṣuvajjyonakrāntisiddhā sodaggatārkāgrā.*

32. The sine of the degrees by which the Sun at midday has risen above the horizon will be the sine of altitude of the Sun at midday. The sine of the degrees by which the Sun is below the zenith at midday will be the midday shadow.

33. Multiply the meridian-sine by the orient-sine and divide by the radius. The square root of the difference between the squares of this result and of the meridian-sine will be the sine of the ecliptic zenith-distance.

The *madhyajyā* or "meridian-sine" is the sine of the zenith-distance of the meridian ecliptic point.

The *udayajyā* or "orient-sine" is the sine of the amplitude of that point of the ecliptic which is on the horizon.

The sine of the ecliptic zenith-distance of that point of the ecliptic which has the greatest altitude (nonagesimal point) is called the *dṛkkṣepajyā*.[3]

[1] Cf. *Sūryasiddhānta,* III, 25–26 n.; Brahmagupta, III, 52; *Pañcasiddhāntikā,* IV, 32–3, 35 n.

[2] *JRAS,* 1865, p. 402.

[3] Cf. *Sūryasiddhānta,* V, 4–6; *Pañcasiddhāntikā,* IX, 19–20 and note; Lalla, *Sūryagrahaṇādhikāra,* 5–6; Kaye, *op. cit.,* pp. 76–77; *BCMS,* XIX (1928), 36.

Brahmagupta (XI, 29–30) criticizes this stanza as follows:

vitribhalagne dṛkkṣepamaṇḍalaṁ tadapamaṇḍalayutau jyā |
madhyā dṛkkṣepajyā nāryabhaṭoktānayā tulyā ||
dṛkkṣepajyāto 'sat tannāśād avanater nāśaḥ |
avanatināśād grāsasyonādhikatā ravigrahaṇe. ||

34. The square root of the difference of the squares of the sines of the ecliptic zenith-distance and of the zenith-distance is the sine of the ecliptic-altitude.

kuvaśāt kṣitije svā dṛk chāyā bhūvyāsārdhaṁ nabhomadhyāt.

The sine of the altitude of the nonagesimal point of the ecliptic is called the *dṛggatijyā*.

Dṛk is equivalent to *dṛgjyā* the sine of the zenith-distance of any planet.[1]

This stanza is criticized by Brahmagupta (XI, 27):

dṛkkṣepajyā bāhur dṛgjyā karṇo 'nayoḥ kṛtiviśeṣāt |
mūlaṁ dṛgnatijīvā saṁsthānam ayuktam etad api. ||

The construction of the second part of the stanza and the exact meaning of *dṛk* and *chāyā* are not clear to me. It seems to mean that when the sine of the zenith-distance is equal to the radius the greatest parallax (horizontal parallax) is equal to the radius of the Earth. *Kuvaśāt* ("because of the Earth") seems to indicate that parallax is due to the fact that we are situated on its surface and not at its center, and that parallax, therefore, is the difference between the positions of an object as seen from the center and from the surface of the Earth.

[1] Cf. *Sūryasiddhānta*, V, 6; Lalla, *Sūryagrahaṇa*, 6; *Pañcasiddhāntikā*, IX, 21 and p. 60; Bhāskara, *Gaṇitādhyāya*, *Sūryagrahaṇa*, II, 5–6 (and *Vāsanābhāṣya*); *BCMS*, XIX (1928), 36–37.

Parameśvara's explanation is as follows:

Dṛgbhedahetubhūtā svacchāyā dṛgjyā vā svadṛggatijyā vā dṛkkṣepajyā vety arthaḥ. sā yadi kṣitije bhavati nabhomadhyāt kṣitijāntā bhavati vyāsārdhatulyā bhavatīty arthas tadā kuvaśād bhūmivaśān niṣpanno dṛgbhedo vyāsārdhaṁ bhavati bhūvyāsārdhatulyaṁ dṛgbhedayojanam ity arthaḥ. antarāle 'nupātāt kalpyam.

Sukumar Ranjan Das[1] states that there is no reference to parallax in Āryabhaṭa. If Parameśvara is correct in interpreting the second part of the rule as giving *yojanas* of *dṛgbheda* (parallax), we must ascribe to Āryabhaṭa the knowledge of parallax, even though no rules are given for its calculation at intermediate points. It is hard to see what else the "radius of the Earth" can refer to when given immediately after rules for finding the *dṛkkṣepajyā* and the *dṛggatijyā* (cf. Brahmagupta, XXI, 64–65, and Bhāskara, *Golādhyāya, Grahaṇavāsanā,* 11–17), especially since parallax was well known to the old *Sūryasiddhānta* which antedated Āryabhaṭa.[2]

It seems to me that the passage is probably to be interpreted in the light of Brahmagupta, XXI, 64–65:

dṛśyādṛśyaṁ dṛggolārdhaṁ bhūvyāsadalavihīnayutam |
draṣṭā bhūgolopari yatas tato lambanāvanatī ||
kṣitije bhūdalaliptāḥ kakṣāyāṁ dṛṅnatir nabhomadhyāt |
avanatiliptā yāmyottarā ravigrahavad anyatra. ||[3]

35. The sine of latitude multiplied by the sine of celestial latitude and divided by the sine of co-latitude is minus and plus to the Moon when it is north of the ecliptic depending on

[1] "Parallax in Hindu Astronomy," *BCMS*, XIX (1928), 29–42.

[2] Cf. *Pañcasiddhāntikā*, p. 60.

[3] See also Lalla, *Madhyagativāsanā*, 23–28.

whether it is in the Eastern or Western hemisphere, plus and minus when it is south of the ecliptic under the same circumstances.

This stanza and the next give the calculation called *dṛkkarman*, an operation for determining the point on the ecliptic to which a planet having a given latitude will be referred by a secondary to the prime vertical. It has been called "operation for apparent longitude" and falls into two parts, namely, the "operation for latitude" (*ākṣadṛkkarman*) treated in this stanza and the "operation for ecliptic-deviation" (*āyanadṛkkarman*) treated in the following stanza.[1]

The stanza is criticized by Brahmagupta (XI, 34):

vikṣepaguṇākṣajyā lambakabhaktā grahe dhanam ṛṇaṁ yat |
uktam udayāstamayayor na pratighaṭikaṁ yatas tad asat. ||

Brahmagupta, X, 13–14 gives a general criticism of Āryabhaṭa's *dṛkkarman*, followed by an exposition of his own method.

36. Multiply the versed sine (of the Moon) by the celestial latitude and by the (greatest) declination, and divide by the square of the radius. The result is minus or plus to the Moon when it is in the northern *ayana* depending on whether its celestial latitude is north or south, and plus or minus when it is in the southern *ayana* under the same conditions.

Parameśvara explains *utkramaṇam* by *koṭyā utkramajyā*.

The *ayanas* are the northern and southern paths of the Sun from solstice to solstice.[2]

[1] Cf. *Sūryasiddhānta*, VII, 8–9 n.; Kaye, *op. cit.*, pp. 78–79.

[2] Cf. *Sūryasiddhānta*, VII, 10 n.; Lalla, *Madhyagativāsanā*, 47–48.

Criticized by Brahmagupta (XI, 35):

trijyākṛtibhaktā vikṣepāpakramaguṇotkramajyendoḥ |
ayanānte yad ṛṇadhanaṁ tat tasyādau tato 'sat tat. ||

37. The Moon consists of water, the Sun of fire, the Earth of earth, and the Earth's shadow of darkness. The Moon obscures the Sun and the great shadow of the Earth obscures the Moon.[1]

Brahmagupta (XI, 9) remarks:

āryabhaṭo jānāti grahāṣṭagatiṁ yad uktavāṁs tad asat |
rāhukṛtaṁ na grahaṇaṁ tatpāto nāṣṭamo rāhuḥ. ||

There is no such statement in our text and Brahmagupta himself (XXI, 43–48) ascribes eclipses to Rāhu.

38. When at the end of the true lunar month the Moon, being near the node, enters the Sun, or when at the end of the half-month the Moon enters the shadow of the Earth that is the middle of the eclipse which occurs sometimes before and sometimes after the exact end of the lunar month or half-month.

Paremeśvara remarks, *sphuṭaśaśimāsānte lambanasaṁskṛte 'māvāsyāntakāle.* He also takes the words *adhikonam* as meaning "middle of the eclipse which lasts for a longer or shorter time," but gives as an alternate explanation offered by some the foregoing translation.[2]

39. Multiply the distance between the Earth and the Sun by the diameter of the Earth and divide by the difference between the diameters of the Earth and the Sun. The result will be the length of the shadow of the Earth (measured) from the diameter of the Earth.

[1] Cf. *Sūryasiddhānta,* IV, 9; Lalla, *Madhyagativāsanā,* 29, 34.

[2] Cf. *Sūryasiddhānta,* IV, 6, 16; Lalla, *Candragrahaṇa,* 10; Alberuni, II, 111.

The last clause seems to indicate that the measurement is to be reckoned from the center of the Earth.[1]

40. The difference between the length of the Earth's shadow and the distance of the Moon from the Earth multiplied by the diameter of the Earth and divided by the length of the Earth's shadow is the diameter of the Earth's shadow (in the orbit of the Moon).[2]

41. Subtract the square of the celestial latitude of the Moon from the square of half the sum (of the diameters of the Sun and Moon or of the Moon and the shadow). The square root of the remainder is known as the *sthityardha.* From this the time is calculated by means of the daily motions of the Sun and Moon.

The *sthityardha* is half of the time from first to last contact.[3]

42. Subtract the radius of the Moon from the radius of the Earth's shadow. Subtract from the square of the remainder the square of the celestial latitude. The square root of this remainder will be the *vimardārdha.*

The *vimardārdha* denotes half of the time of total obscuration.[4]

43. Subtract the radius of the Moon from the radius of the Earth's shadow. Subtract this remainder from the celestial latitude. The remainder is the part of the Moon which is not eclipsed.

44. Subtract the given time from half of the duration of the obscuration. Add this to the square of the celestial latitude. Take

[1] Cf. Brahmagupta, XXIII, 8.

[2] Cf. Brahmagupta, XXIII, 9.

[3] Cf. *Pañcasiddhāntikā*, VI, 3 and X, 2b–3; *Sūryasiddhānta*, IV, 12–13; Brahmagupta, IV, 8.

[4] Cf. *Sūryasiddhānta*, IV, 13; *Pañcasiddhāntikā*, X, 7; Brahmagupta, IV, 8.

the square root. Subtract this from half the sum of the diameters. The remainder will be the obscuration at the given time.[1]

The first sentence ought to be: "Subtract the *koṭi* of the given time from the *koṭi* of the *sthityardha*. Square this."

45. The sine of the latitude multiplied by the sine of the hour-angle and divided by the radius is the deflection due to latitude. It is south.

sthityardhāc cārkendos trirāśisahitāyanāt sparśe.

For the difficulty of the stanza and the gap in the commentary of Parameśvara see the Preface to Kern's edition (pp. v–vi) with the references to Bhāskara.

"Hour-angle" is expressed by *madhyāhnāt krama* (*guṇitaḥ*). "Deflection due to latitude" seems to be the meaning of *dik*.

The first part deals with the *ākṣavalana* or "deflection due to latitude." According to Parameśvara, it is south in the Eastern and north in the Western hemisphere. The other books give just the opposite.

Parameśvara remarks, *etad akṣavalanaṁ sthityardhāc ca. sthityardhaśabdena tanmūlabhūto vikṣepa ucyate.*

Parameśvara also remarks, *ayanaśabdenāpakrama ucyate. trirāśisahitād arkāc candrāc ca niṣpanno 'pakramo 'pi tayor arkendor valanam bhavati.*

Parameśvara explains *sparśe* as *sparśa iti grahaṇa ity evārthataḥ.*

However the second part of the stanza is to be

[1] Cf. *Sūryasiddhānta*, IV, 18–20; *Pañcasiddhāntikā*, X, 5–6; Brahmagupta, IV, 11–12.

translated it must deal with the so-called *āyanavalana* or "deflection due to the deviation of the ecliptic from the equator."

Both *valanas* ("deflection of the ecliptic") were employed in the projection of eclipses.[1]

46. At the beginning of an eclipse the Moon is *dhūmra*, when half obscured it is *kṛṣṇa*, when completely obscured it is *kapila*, at the middle of an eclipse it is *kṛṣṇatāmra*.[2]

47. When the Moon eclipses the Sun even though an eighth part of the Sun is covered this is not preceptible because of the brightness of the Sun and the transparency of the Moon.[3]

48. The Sun has been calculated from the conjunction of the Earth and the Sun, the Moon from the conjunction of the Sun and Moon, and all the other planets from the conjunctions of the planets and the Moon.[4]

49. By the grace of God the precious sunken jewel of true knowledge has been rescued by me, by means of the boat of my own knowledge, from the ocean which consists of true and false knowledge.[5]

50. He who disparages this universally true science of astronomy, which formerly was revealed by Svayambhū, and is now described by me in this Āryabhaṭīya, loses his good deeds and his long life.[6]

Read *pratikuñcuko*.

[1] Cf. Brahmagupta, IV, 16–17 and XXI, 66; Lalla, *Candragrahaṇādhikāra*, 23, 25; *Sūryasiddhānta*, IV, 24–25: "From the position of the eclipsed body increased by three signs calculate the degrees of declination."

See Brennand, *Hindu Astronomy*, pp. 280–83; Kaye, *Hindu Astronomy*, pp. 77–78.

[2] Cf. *Sūryasiddhānta*, VI, 23; Lalla, *Candragrahaṇādhikāra*, 36; Brahmagupta, IV, 19.

[3] Cf. *Sūryasiddhānta*, VI, 13.

[4] Cf. *BCMS*, XII (1920–21), 183.

[5] Cf. *ibid.*, p. 187.

[6] Cf. *JRAS*, 1911, p. 114.

GENERAL INDEX

[Including the most important Sanskrit proper names]

SANSKRIT INDEX

[PRINTED IN U S A]

CPSIA information can be obtained at www.ICGtesting.com
Printed in the USA
LVOW101754040412

276176LV00001B/376/A